SCHAUM'S OUTLINE OF

THEORY AND PROBLEMS

of

INTRODUCTION
to
ENGINEERING
CALCULATIONS
for First-Year Students

by

BYRON S. GOTTFRIED, Ph. D.

Professor of Industrial Engineering
Systems Management Engineering and Operations Research
University of Pittsburgh

SCHAUM'S OUTLINE SERIES

McGRAW-HILL BOOK COMPANY

New York   St. Louis   San Francisco   Auckland   Bogotá   Düsseldorf   Johannesburg
London   Madrid   Mexico   Montreal   New Delhi   Panama   Paris
São Paulo   Singapore   Sydney   Tokyo   Toronto

To Sharon, Gail, and Susan

BYRON S. GOTTFRIED is a Professor of Industrial Engineering at the University of Pittsburgh, where he has been a member of the faculty since 1970. His major area of interest is the use of computers for solving complex technical and business problems. Dr. Gottfried received his Ph.D. in Chemical Engineering from Case-Western Reserve University, and has worked for Gulf Oil Corporation, Westinghouse Electric Corporation, and the National Aeronautics and Space Administration. He is the author of several textbooks, as well as Schaum's Outline of Programming with Basic.

Schaum's Outline of Theory and Problems of
INTRODUCTION TO ENGINEERING CALCULATIONS

3 4 5 6 7 8 9 10 11 12 13 14 15 SH SH 8 7 6 5 4 3 2 1

Sponsoring Editor, David Beckwith
Editing Supervisor, Denise Schanck
Production Manager, Nick Monti

Library of Congress Cataloging in Publication Data

Gottfried, Byron S., date.
   Schaum's outline of introduction to engineering calculations.

   (Schaum's outline series)
   Includes index.
   1. Engineering mathematics.   I.   Title.
II.   Title: Elementary engineering calculations.
TA330.G67   510'.2'462   78-27293
ISBN 0-07-023837-5

Preface

There are certain basic computational skills that should be mastered by all beginning engineering students. These include a familiarity with units and units conversions, including SI units; simple graphical and numerical techniques for displaying data, analyzing data, and solving rudimentary computational problems; a knowledge of elementary engineering economics; and insight into the uses and limitations of electronic computational devices. Most engineering educators assume that their students have been formally exposed to this material at some early point in their academic careers, though I have found that this assumption is frequently incorrect. In reality, many freshman engineering students are plunged directly into the more sophisticated aspects of engineering analysis without an adequate background in fundamental computational procedures.

This book is intended to remedy these deficiencies by presenting a number of simple, important techniques of contemporary engineering analysis. Much of the material consists of traditional topics that have formed the cornerstone of engineering analysis for many years. These topics include computational precision and numerical approximation (Chapter 1), units systems and units conversions (Chapter 2), the graphical representation of data (Chapter 3), elementary data reduction techniques (Chapter 4), and graphical as well as numerical techniques for solving algebraic equations and evaluating integrals (Chapter 5).

With the inclusion of additional material of a more timely nature, these traditional topics are placed in the context of present-day professional practice. This additional material includes a discussion of the electronic calculator (Chapter 1), standard international units (Chapter 2), and a rather general introduction to computers (Chapter 7), which emphasizes organization and flowcharting of algorithmic processes without referring to any particular programming language. This last topic provides an excellent background for a subsequent course in computer programming. Some introductory material on engineering economics is also included (Chapter 6), in order to stress the importance of economic criteria in the development of new technology. The international metric system (SI units) is used exclusively throughout the book, except in Chapter 2, which considers both the international metric system and the American engineering system of units.

The book contains a large number of examples and solved problems, in addition to an extensive set of supplementary problems at the end of each chapter. Thus the book can be used as a supplement to textbooks in this area, as a textbook in its own right, or as an effective self-study guide. Its use as a supplementary text in conjunction with the more traditional types of introductory engineering textbooks is particularly attractive.

I wish to express my gratitude to Mary Catherine Kappauf for meticulously checking all the problems and examples, and to Freda Stephen for patiently and diligently typing the manuscript. Also, I wish to thank my students and colleagues who examined early portions of the manuscript and made many helpful suggestions. Finally, I must thank the editorial staff of the Schaum's Outline Series for their excellent cooperation and support.

Byron S. Gottfried

Contents

CONTENTS

Chapter 1

Arithmetic Calculations

1.1 SCIENTIFIC NOTATION

In *scientific notation* a quantity is expressed as a number between 1 and 9.999 999 ... , multiplied by 10 to an appropriate power. The numerical quantity containing the decimal point is called the *mantissa*, and the power of 10 is called the *exponent*. The conventional form of the number is found by shifting the decimal point in the mantissa to the right (for a positive exponent) or to the left (for a negative exponent).

Example 1.1. (*a*) The number 129 can be expressed in scientific notation as 1.29×10^2. Thus the term 10^2 causes the decimal point to be shifted two places to the right. Another way to express this is to write

$$1.29 \times 10^2 = 1.29 \times 100 = 129$$

(*b*) The number 0.005 15 can be written in scientific notation as 5.15×10^{-3}. The term 10^{-3} causes the decimal point to be shifted three places to the left. Therefore,

$$5.15 \times 10^{-3} = 5.15 \times 0.001 = 0.005\ 15$$

Scientific notation is particularly useful when writing numbers that are very large or very small.

Example 1.2. The mass of one electron is

$$0.000\ 000\ 000\ 000\ 000\ 000\ 000\ 000\ 000\ 000\ 910\ 86 \text{ kg}$$

The superiority of the scientific notation,

$$9.1086 \times 10^{-31} \text{ kg}$$

is readily apparent in this case.

Example 1.3.

Conventional Notation	Scientific Notation
1	1×10^0
-1	-1×10^0
0.001	1×10^{-3}
1000	1×10^3
123 456.789	$1.234\ 567\ 89 \times 10^5$
$-123\ 456.789$	$-1.234\ 567\ 89 \times 10^5$
0.000 012 345 678 9	$1.234\ 567\ 89 \times 10^{-5}$
$-0.000\ 012\ 345\ 678\ 9$	$-1.234\ 567\ 89 \times 10^{-5}$
$-1.234\ 567\ 89$	$-1.234\ 567\ 89 \times 10^0$

1.2 NORMALIZED FLOATING POINT NOTATION

Computer systems also use scientific notation, but in a somewhat different form. A number is represented by a mantissa whose value ranges between 0.1 and 0.999 999 ... , followed by an appropriate power of 10. Furthermore, the power of 10 is usually shown as $E+n$ or $E-n$, where $+n$ or $-n$ represents the exponent.

1

Example 1.4.

Conventional Notation	Normalized Floating Point Notation
129	0.129E+3
0.005 15	0.515E−2
0.1	0.1E+0
−1	−0.1E+1
0.001	0.1E−2
1000	0.1E+4
123 456.789	0.123 456 789E+6
−123 456.789	−0.123 456 789E+6
0.000 012 345 678 9	0.123 456 789E−4
−0.000 012 345 678 9	−0.123 456 789E−4
−1.234 567 89	−0.123 456 789E+1

In some computer systems the sign is not shown when the exponent is positive. Also, some computer systems always use a two-digit exponent.

Example 1.5. In some computer systems the number 129 will appear as 0.129E 3; in others, as 0.129E 03. Similarly, the number 0.005 15 might appear as 0.515E−2 or as 0.515E−02.

1.3 SIGNIFICANT FIGURES

The *precision* of a numerical quantity is determined by the number of *significant figures* in the quantity. If the quantity is written in normalized floating point notation, then the number of significant figures is the number of digits to the right of the decimal point.

Example 1.6. (*a*) The quantity 129 consists of three significant figures, since the normalized floating point number 0.129×10^3 (or 0.129E+3) contains three digits to the right of the decimal point. (*b*) The quantity 0.005 15 consists of three significant figures, since 0.515×10^{-2} (or 0.515E−2) contains three digits to the right of the decimal point. Notice that the two zeros that appear immediately after the decimal point in 0.005 15 *are not* considered to be significant figures.

Zeros that appear within a normalized floating point number (*imbedded zeros*) and zeros that appear at the end of the number (*trailing zeros*) are included as significant figures. However, the zero preceding the decimal point (*leading zero*) is *not* considered a significant figure.

Example 1.7. The number 408.7 consists of four significant figures, since 0.4087×10^3 (or 0.4087E+3) contains four digits (including the zero) to the right of the decimal point. Similarly, the number 408.700 consists of six significant figures, since $0.408 700 \times 10^3$ (or 0.408 700E+3) contains six digits (including three zeros) to the right of the decimal point.

When an integer quantity ending in zero is written in conventional notation, it is not clear whether or not the trailing zeros are significant. This ambiguity can be eliminated by writing the number in normalized floating point notation.

Example 1.8. The quantity 1000 contains at least one, and perhaps as many as four, significant figures, but the exact number of significant figures is not specified. In normalized floating point form, however, the quantity can be written as 0.1×10^4 (or 0.1E+4), which clearly contains only one significant figure, or as 0.1000×10^4 (or 0.1000E+4), which contains four significant figures. Intermediate levels of precision can also be specified by writing 0.10×10^4 or 0.100×10^4.

The *most significant digit* in a number is the leftmost nonzero digit. The *least significant digit* is the rightmost significant digit (which may be zero).

Example 1.9. In the number 408.7, the most significant digit is 4 and the least significant digit is 7. Similarly, in the number 408.700, the most significant digit is 4 and the least significant digit is 0.

1.4   ROUNDING

In many situations it is desirable to reduce the number of significant figures in a quantity by eliminating the least significant digit. This is usually accomplished by *rounding*. The rules are as follows.

1. If the least significant digit is less than 5, then it is simply dropped. The remaining digits are unchanged. (This is known as *rounding down*.)

Example 1.10. The quantity 0.432 becomes 0.43 when rounded to two significant figures.

2. If the least significant digit is greater than 5, then the next significant digit is increased by 1. (This is known as *rounding up*.)

Example 1.11. The quantity 0.5066 becomes 0.507 when rounded to three significant figures.

3. If the least significant digit is exactly 5, then the next significant digit is increased by 1 if it is odd. If the next significant digit is even, however, it remains unchanged. (This is known as the *odd-add rule*.)

Example 1.12. The quantity 0.2075 becomes 0.208 when rounded to three significant figures (because 7 is an odd digit). However, the quantity 0.2045 is written as 0.204 when rounded to three significant figures (because 4 is even).

The odd-add rule is somewhat arbitrary, but there are two advantages associated with it. First, it is consistent and therefore eliminates ambiguities. Secondly, the errors that result from its use tend to cancel one another when the rule is employed repeatedly in a lengthy calculation.

Multiple-digit rounding is carried out in the same manner. For example, if two significant figures are to be dropped, then round down if the last two digits are less than 50, round up if they are greater than 50, and apply the odd-add rule if they equal 50.

Example 1.13. Shown below are several four-digit numbers that are rounded to two significant figures.

0.2448	becomes	0.24	(round down)
0.2453	becomes	0.25	(round up)
0.2450	becomes	0.24	(odd-add rule)
0.2550	becomes	0.26	(odd-add rule)
0.2548	becomes	0.25	(round down)

Notice that the last number would incorrectly be rounded up to 0.26 if the rounding were carried out one digit at a time. In other words, 0.2548 would first be rounded up to 0.255, which then becomes 0.26 because of the odd-add rule. Thus multiple-digit rounding should not be carried out one digit at a time.

1.5   TRUNCATING

If significant figures are dropped without ever rounding up, the number is said to be *truncated* (or *chopped*). Clearly, the truncated value of a number will always be less than or equal to the rounded value of the number.

1.6 ADDITION AND SUBTRACTION

In order to perform addition or subtraction with numbers that are expressed in scientific notation (or normalized floating point notation), each of the numbers must have the same exponent (i.e. the same power of 10).

Example 1.14. Calculate the sum of 0.5853×10^{-2} and 0.4011×10^{-3}.
The second number can be written as $0.040\,11\times10^{-2}$. Hence, the problem becomes

$$\begin{array}{r} 0.585\,3\ \times10^{-2} \\ +0.040\,11\times10^{-2} \\ \hline 0.625\,41\times10^{-2} \end{array}$$

Rounding to four significant figures, the sum is 0.6254×10^{-2}.

The sum or difference of two numbers is normally expressed with the same degree of precision as the number with the fewer significant figures.

Example 1.15. Calculate the difference of 0.75×10^{-3} and 0.127×10^{-4}.
The second number can be written as 0.0127×10^{-3}. Therefore, the problem is

$$\begin{array}{r} 0.75\ \ \times10^{-3} \\ -0.0127\times10^{-3} \\ \hline 0.7373\times10^{-3} \end{array}$$

Since the first number has only two significant figures, the final result will also be expressed in terms of two significant figures. The desired difference, then, is 0.74×10^{-3}.

1.7 MULTIPLICATION AND DIVISION

When multiplying or dividing numbers that are written in scientific notation (or normalized floating point notation), it is necessary to multiply or divide the mantissas and to add or subtract the corresponding exponents. Specifically, if $n_1 = a_1\times10^{e_1}$ and $n_2 = a_2\times10^{e_2}$, then

$$n_1\times n_2 = (a_1\times a_2)\times10^{e_1+e_2} \qquad \text{and} \qquad \frac{n_1}{n_2} = \frac{a_1}{a_2}\times10^{e_1-e_2}$$

Example 1.16. Obtain the product of 0.483×10^{-3} and 0.208×10^2, to three significant figures.
The problem can be expressed as

$$(0.483\times0.208)\times10^{-3+2} = 0.100\,464\times10^{-1}$$

Therefore the product, to three significant figures, is 0.100×10^{-1}.

Example 1.17. Divide 0.483×10^{-3} by 0.208×10^2. Express the quotient in terms of three significant figures.
This problem can be written as

$$\frac{0.483}{0.208}\times10^{-3-2} = 2.322\,115\,4\ldots\times10^{-5} = 0.232\,211\,54\ldots\times10^{-4}$$

Therefore the quotient, to three significant figures, is 0.232×10^{-4}.

The product or quotient of two numbers is normally written with the same degree of precision as the number with the fewer significant figures.

Example 1.18. Evaluate the formula $w = xy/z$ where $x = 0.645\times10^{-2}$, $y = 0.27\times10^{-3}$ and $z = 0.1877\times10^4$.

The problem becomes

$$w = \frac{(0.645 \times 10^{-2})(0.27 \times 10^{-3})}{0.1877 \times 10^4} = \frac{(0.645)(0.27)}{0.1877} \times 10^{-2-3-4} = 0.93 \times 10^{-9}$$

Notice that w is expressed in terms of two significant figures because y contains only two significant figures.

1.8   EXPONENTIATION

If a number written in scientific notation is raised to a power, then the mantissa is raised to that power and the exponent is multiplied by the given power.   In other words, if the original number is expressed as $n = a \times 10^e$,   then

$$n^k = a^k \times 10^{ke}$$

This operation is known as *exponentiation*.   It is valid both for integer and fractional values of k (and combinations thereof).

Example 1.19.   Calculate the square and the square root of 0.6538×10^{-6}.
The square can be expressed as

$$(0.6538 \times 10^{-6})^2 = (0.6538)^2 \times 10^{(2)(-6)} = 0.4275 \times 10^{-12}$$

Similarly, the square root can be obtained by writing

$$(0.6538 \times 10^{-6})^{0.5} = (0.6538)^{0.5} \times 10^{(0.5)(-6)} = 0.8086 \times 10^{-3}$$

When raising a number to a fractional power, it may be desirable to increase or decrease the original exponent (by shifting the location of the decimal point within the mantissa) so that the final exponent will be an integer.

Example 1.20.   Calculate the cube root of 0.8375×10^{-5}.
If the given exponent, -5, is multiplied by $1/3$, the resulting exponent will not be an integer.   However, if we write the given number as 8.375×10^{-6}, the answer is obtained as

$$(8.375 \times 10^{-6})^{1/3} = (8.375)^{1/3} \times 10^{(1/3)(-6)} = 2.031 \times 10^{-2} = 0.2031 \times 10^{-1}$$

1.9   COMBINED OPERATIONS

In more complicated calculations, the intermediate results should be expressed as precisely as possible (within reason).   The final answer is then rounded so that it does not contain more significant figures than any of the numbers in the original data.

Example 1.21.   Using an 8-digit calculator, evaluate the formula

$$x = \frac{-b + \sqrt{b^2 - 4ac}}{2a}$$

for the following data:

$$a = 0.1406 \times 10^{-3} \qquad b = 0.4774 \times 10^0 \qquad c = 0.4052 \times 10^3$$

Substituting the data into the formula, we obtain

$$x = \frac{-0.4774 + \sqrt{(0.4774)^2 - (4)(0.1406 \times 10^{-3})(0.4052 \times 10^3)}}{(2)(0.1406 \times 10^{-3})}$$

$$= \frac{-0.4774 + \sqrt{0.227\,910\,76 - 0.227\,884\,48}}{0.2812 \times 10^{-3}} = \frac{-0.4774 + \sqrt{0.000\,026\,28}}{0.2812 \times 10^{-3}}$$

$$= \frac{-0.4774 + 0.005\,126\,402\,2}{0.2812 \times 10^{-3}} = \frac{-0.472\,273\,60}{0.2812 \times 10^{-3}} = -1.679\,493\,6 \times 10^3$$

Rounding to four significant figures, we obtain the desired result:

$$x = -1.679 \times 10^3 = -0.1679 \times 10^4$$

Notice that eight significant figures were carried whenever possible, until the very end of the calculation. If we had rounded to four significant figures in the intermediate calculations, we would have obtained

$$x = \frac{-0.4774 + \sqrt{0.2279 - 0.2279}}{0.2812 \times 10^{-3}} = \frac{-0.4774}{0.2812 \times 10^{-3}} = -1.697\,724\,0 \times 10^3$$

Thus the final answer would have been $x = -0.1698 \times 10^4$,   which is accurate to only two significant figures.

1.10   APPROXIMATE CALCULATIONS

Many technical calculations need not be carried out to a high degree of precision.   Sometimes we require only a rough answer, either to get some overall "feel" for a problem or to check a result obtained by more sophisticated means.

To find an approximate solution to a problem that involves addition, subtraction, multiplication, and division write each number in scientific notation, retaining only two significant figures.   The arithmetic operations can then be carried out mentally to a reasonable degree of accuracy.   Errors that result from this procedure tend to cancel one another.

Example 1.22.   Consider the formula

$$w = \frac{a}{bc} - \frac{xy}{z}$$

Obtain an approximate value for w, given

$a = 27.86$	$x = 0.006\,674$
$b = 0.043\,51$	$y = 16.498$
$c = 88.266$	$z = 0.054\,707$

The problem can be written as (the symbol $\approx$ means "equals approximately")

$$w \approx \frac{2.8 \times 10^1}{(4.4 \times 10^{-2})(0.88 \times 10^2)} - \frac{(0.67 \times 10^{-2})(1.6 \times 10^1)}{0.55 \times 10^{-1}}$$

$$\approx \frac{28}{4} - \frac{0.11}{0.055} \approx 7 - 2 = 5$$

Thus, using mental arithmetic, we obtain $w \approx 5$.   (The exact answer, to four significant figures, is 5.242.)

Exponentiation operations are more difficult to approximate mentally.   Problems involving a fractional (decimal) exponent can be especially troublesome, since the magnitude of the final result may not be readily apparent.   If the exponent is a positive fraction, however, then the result of exponentiation will lie somewhere between 1 and the original number.   (Recall that any nonzero number raised to the zeroth power is 1; i.e., $n^0 = 1$.)   Note that square roots, cube roots, etc., fall into this category.

An operation involving a negative fractional exponent can always be written as the reciprocal of an operation involving a positive fractional exponent.

Example 1.23.   Estimate the value of $(45)^{-0.3}$.   (At first glance, the reader might have no idea what the magnitude of the expected answer will be!)

Let us rewrite the problem as $1/(45)^{0.3}$.   We know that the value of the denominator will be between 1 and 45.   Moreover, we know that $(27)^{1/3} = 3$ and $(64)^{1/3} = 4$. Hence we can write

$$(45)^{0.3} \approx 3.5 \qquad \text{and} \qquad (45)^{-0.3} \approx \frac{1}{3.5} \approx 0.3$$

(The correct answer is 0.3192.)

If the numbers used in a calculation are not known accurately, then a precise, detailed solution will not be meaningful.   Approximate solutions are particularly appropriate in such situations.

Example 1.24.   A large computer is able to carry out a basic arithmetic operation in about 2 nanoseconds $(2\times10^{-9}$ s).   Computer time costs \$600 an hour.   How much would it cost to process a payroll for 1000 employees, assuming that the records for each employee require about 50 arithmetic operations?
    The total number of arithmetic operations needed to process the entire payroll will be

$$50\times1000 = 50\,000 = 5\times10^4$$

so that the required computer time will be

$$(5\times10^4)(2\times10^{-9}) = 10\times10^{-5} = 10^{-4} \text{ s}$$

Since computer time costs \$600 per hour, the total cost (in dollars) of processing the payroll will be

$$\left(\frac{600}{3600}\right)(10^{-4}) = \frac{1}{6}\times10^{-4} \approx 0.17\times10^{-4}$$

In other words, the entire payroll can be processed for only about 0.002 cents!
    From this simple calculation we have obtained a very significant conclusion—namely, that the cost of using a large computer for a typical commercial application is extremely low, despite the high hourly charge.   Moreover, this conclusion will be correct even if the estimated data (the assumed number of arithmetic calculations required for each employee) is considerably in error.

1.11   ALGEBRAIC AND TRIGONOMETRIC FUNCTIONS

The exponential, logarithmic and trigonometric functions appear frequently in technical calculations.   These functions can best be evaluated with an electronic calculator or a set of tables.
    The evaluation of a function can be thought of as an operation that transforms a given number into a new number.   The given number is called an *argument* of the function.

Example 1.25.   Suppose that we wanted to evaluate the formulas

$$y = e^x \qquad y = \log x \qquad y = \sin x$$

for a given value of x.   Each formula involves the evaluation of a single function (namely, the exponential function, the logarithmic function and the sine function).   In each case the argument, x, is transformed into a new quantity, y.

The Exponential Function

In calculus it is shown that the quantity $(1+m^{-1})^m$ approaches a constant value of $2.718\,281\,8\ldots$ as m becomes very large.   This quantity, which is represented by the letter e, is the base of the *natural* (or *Naperian*) system of logarithms.

Many technical problems require exponentiation of the quantity e.   Hence the function e^x is known as the *exponential function*.   A distinction is sometimes made between the *positive exponential function*, for which $x \geqslant 0$, and the *negative exponential function*, where $x \leqslant 0$.   The latter is often written as $y = e^{-x}$, $x \geqslant 0$.

Figure 1-1 shows a plot of the positive and negative exponential functions.   Notice that the positive exponential function approaches infinity as x becomes large, whereas the negative exponential function approaches zero.   Also, note that both exponential functions approach 1 as x approaches zero (since $e^0 = 1$).   The important characteristics of the exponential function are summarized in Table 1-1, page 13.

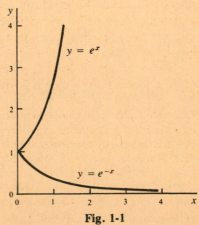

Fig. 1-1

Example 1.26. The current in an electrical circuit can be represented by the equation

$$y = ae^{-bt}$$

where y represents the current (milliamperes), t represents time (seconds), and a and b are positive constants (whose units are milliamperes and inverse seconds, respectively). If $a = 2$ and $b = 0.5$, determine the current after 3 s. What fraction is this of the initial current (that at $t = 0$)?

The current after 3 s can be obtained by writing

$$y = 2e^{-(0.5)(3)} = 2e^{-1.5}$$

The exponential function can easily be evaluated with an electronic calculator. Thus $e^{-1.5} = 0.223\ 130\ 16$, and

$$y = 2(0.223\ 130\ 16) = 0.446\ 260\ 32 \approx 0.45 \text{ mA}$$

The initial current can easily be obtained as

$$y = 2e^{-(0.5)(0)} = 2e^0 = 2 \text{ mA}$$

Therefore, the fraction of the initial current remaining after 3 s is $0.45/2 = 0.225$.

Notice that the current becomes smaller and smaller as t increases, until it essentially vanishes when t is very large. This type of behavior is known as *exponential decay*.

If x is small (in magnitude), the exponential function can be approximated as

$$e^x \approx 1 + x$$

This approximation is valid to at least four decimal places for $|x| \le 0.01$, and at least two decimal places for $|x| \le 0.1$. The approximation is useful for checking results, or for evaluating the exponential function if a calculator or set of tables is not available.

Example 1.27. A radioactive substance, A, decays exponentially to form a new substance, B, which in turn decays (more slowly) to form a stable substance, C. The sequence of events can be represented schematically as

$$A \to B \to C$$

If we begin with pure A, then the concentration of B at any time t can be determined by the equation

$$y = \frac{c_1}{c_1 - c_2}(e^{-c_2 t} - e^{-c_1 t})$$

where y represents the concentration of B (expressed as a pure number), t represents time (in hours), and c_1 and c_2 (both in h^{-1}) are called *decay constants*. Determine the concentration of B after 1 h for the case where $c_1 = 0.07$ and $c_2 = 0.02$.

Substituting into the above equation, we obtain

$$y = \frac{0.07}{0.07 - 0.02}[e^{-(0.02)(1)} - e^{-(0.07)(1)}] = 1.4[e^{-0.02} - e^{-0.07}]$$

$$\approx 1.4[(1 - 0.02) - (1 - 0.07)] = (1.4)(0.05) = 0.07$$

The correct answer, obtained with an electronic calculator, is

$$y = 1.4(0.980\ 198\ 67 - 0.932\ 393\ 82) \approx 0.0669$$

The important characteristics of the exponential function are summarized in Table 1-1, page 13.

The Logarithmic Functions

Two types of logarithmic functions are used in technical calculations, the *common (base 10) logarithmic function* and the *natural (base e) logarithmic function*. In each case the logarithm is defined as the power to which the base must be raised in order to obtain a desired number. In other words, if $x = 10^y$, then y is the (common) logarithm of x. This is usually written $y = \log_{10} x$, or simply $y = \log x$. Similarly, if $x = e^y$, then y is the natural logarithm of x. This is written as $y = \log_e x$ or, more commonly, $y = \ln x$.

Figure 1-2 contains plots of $\log x$ and $\ln x$. Notice that the logarithmic functions become very large as x increases, and they approach $-\infty$ as x approaches zero. Also, notice that the logarithmic functions are negative for $0 < x < 1$, and that $\log 1 = \ln 1 = 0$. Finally, it should be mentioned that the logarithmic functions are not defined for negative values of x.

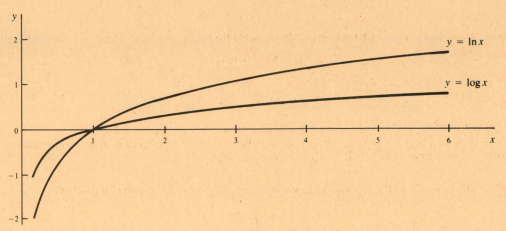

Fig. 1-2

Example 1.28. A cylindrical water tank is emptied by draining the water from a small opening at the bottom. The height of water in the tank and the discharge velocity are related by the equation

$$h = 0.65 \log_{10}(1+8v)$$

where h is the height of water in the tank (in meters) and v is the discharge velocity (meters per second). What height of water corresponds to a discharge velocity of 0.1 m/s?

Substituting the value $v = 0.1$ into the above equation, we obtain

$$h = 0.65 \log(1+8\times0.1) = 0.65 \log 1.8$$

The logarithmic function can be evaluated with an electronic calculator or a table of logarithms, resulting in

$$\log 1.8 = 0.255\,272\,51$$

Hence $h = (0.65)(0.255\,272\,51) \approx 0.166$ m.

Example 1.29. A metal rod at 1300 °C is plunged into a bath of oil whose temperature is maintained at 100 °C. The Celsius temperature at the center of the rod is given by the expression

$$T = 100+1200e^{-0.1t}$$

where t is the time (in s) that the rod remains in the constant-temperature bath. Determine the time required for the center of the rod to cool to 500 °C.

In order to solve this problem, the equation must be rearranged to read

$$\frac{T-100}{1200} = e^{-0.1t}$$

Taking the natural logarithm of each side,

$$\ln\left(\frac{T-100}{1200}\right) = -0.1t \qquad \text{or} \qquad t = -10\ln\left(\frac{T-100}{1200}\right)$$

If we let $T = 500$, we obtain

$$t = -10\ln\left(\frac{500-100}{1200}\right) = -10\ln\frac{1}{3} = -10\,(-1.098\,612\,3) \approx 11 \text{ s}$$

The functions $\log x$ and $\ln x$ are related by

$$\ln x = 2.302\,585\,1 \log x \qquad \text{and} \qquad \log x = 0.434\,294\,48 \ln x$$

Therefore, if one of the logarithmic functions is known, the other can easily be determined. (Notice that the magnitude of $\ln x$ will always be greater than the magnitude of $\log x$.)

If x is close to 1, its natural logarithm can be approximated as

$$\ln x \approx x - 1$$

This approximation is valid to at least two decimal places for $|x-1| \leq 0.01$. The approximation may be useful for checking results, or for estimating the value of logarithms if a calculator or set of tables is not available.

Example 1.30. Estimate the value of the common logarithm of 0.99.

Since the argument (0.99) is close to 1, we can estimate the *natural* log as

$$\ln x \approx x - 1 = 0.99 - 1 = -0.01$$

Hence the common log can be approximated as

$$\log x \approx (0.434\,294\,48)(-0.01) \approx -0.0043$$

The correct answer, to eight decimals, is $-0.004\,364\,81$.

The important characteristics of the logarithmic functions are summarized in Table 1-1, page 13.

The Trigonometric Functions

Of the six trigonometric functions, we will consider only two, the sine and the cosine. These are the most commonly used trigonometric functions. Besides, the remaining four functions can easily be obtained from these two.

Suppose that we are given a right triangle, as shown in Fig. 1-3. For such a triangle the *sine* of the angle θ is defined as the ratio of the side opposite the angle to the hypotenuse. Hence, in Fig. 1-3, $\sin \theta = a/c$. Similarly, the *cosine* of the angle is defined as the ratio of the adjacent side to the hypotenuse. Thus $\cos \theta = b/c$ in Fig. 1-3. From the Pythagorean theorem it is easy to show that the sine and cosine are related by the equation

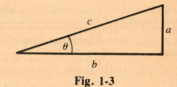

Fig. 1-3

$$\sin^2 \theta + \cos^2 \theta = 1$$

Example 1.31. A force (F) of 8 newtons acts at an angle (θ) of 42° from the horizontal. Determine the horizontal component and the vertical component of this force.

The horizontal and vertical components are given by

$$F_h = F \cos \theta = 8 \cos 42° \qquad F_v = F \sin \theta = 8 \sin 42°$$

The sine and cosine can be evaluated using an electronic calculator (or a set of tables). Thus

$$\sin 42° = 0.669\,130\,61 \qquad \cos 42° = 0.743\,144\,83$$

and

$$F_h = 8\,(0.743\,144\,83) \approx 5.95 \text{ N} \qquad F_v = 8\,(0.669\,130\,61) \approx 5.35 \text{ N}$$

The argument of a trigonometric function can be expressed either in *degrees* or in *radians*, although the latter is more commonly used in technical calculations. Recall that

$$360 \text{ degrees} = 2\pi \text{ radians} \qquad \text{or} \qquad 1 \text{ radian} = 57.3 \text{ degrees}$$

Example 1.32. A long, slender, vertical column is rigidly supported at its base. If a vertical force is applied to the free end, the column will be deflected horizontally from its original vertical position. The deflection at any point y can be determined by the expression

$$d = 2 \sin \left(\frac{3\pi y}{2L} \right)$$

where d = horizontal deflection, mm
        y = vertical distance from base of column, mm
        L = column length, mm
Determine the deflection at a point 1/4 of the way up the column.
    In this case, $y/L = 1/4$.   Hence

$$d = 2 \sin \left(\frac{3\pi}{2} \times \frac{1}{4} \right) = 2 \sin \frac{3\pi}{8} = 2 \sin 1.1781$$

(Note that the argument, 1.1781, is expressed in radians.)   The sine function can now be evaluated using a calculator or a set of tables, resulting in

$$d = 2(0.923\,879\,53) \approx 1.848 \text{ mm}$$

    Figure 1-4 shows plots of $\sin x$ and $\cos x$ against x (in radians).   We see that each function repeats itself for x larger than 2π or x less than 0.   Thus the sine and cosine functions are said to be *periodic*, with period 2π.   Furthermore, we see that the cosine function is symmetric about the y-axis, i.e.

$$\cos(-x) = \cos x$$

and the sine function is antisymmetric, i.e.

$$\sin(-x) = -\sin x$$

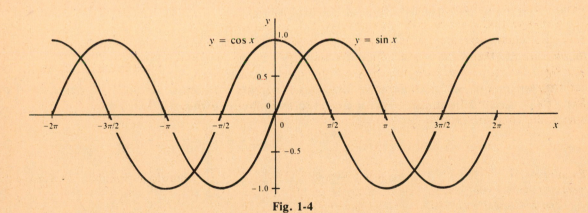

Fig. 1-4

Example 1.33. Use Fig. 1-4 to estimate the value of $y = \sin 50$ (where the argument is expressed in radians).
    Since the sine function is periodic with period 2π, we can write

$$\sin 50 = \sin(50 - 2\pi) = \sin(50 - 4\pi) = \cdots$$

The largest multiple of 2π that does not exceed 50 is

$$7 \times 2\pi = 7 \times 6.283\,185\,3 = 43.982\,297$$

Hence

$$\sin 50 = \sin(50 - 14\pi) = \sin 6.017\,702\,9 \approx \sin(1.9\pi) \approx -0.2$$

The correct answer, to 4 decimals, is $\sin 50 = -0.2624$.

Since the sine and cosine functions are used so often in technical calculations, the reader should memorize the general shape of the two curves shown in Fig. 1-4. In particular, the reader should remember the key values included in Table 1-1, page 13.

Example 1.34. Determine the value of cos (-21π).
Since the cosine function is symmetric and periodic with period 2π, we can write

$$\cos(-21\pi) = \cos 21\pi = \cos 19\pi = \cdots = \cos \pi = -1$$

If x is small (in magnitude), then the sine and cosine functions can be approximated by

$$\sin x \approx x \qquad \cos x \approx 1 - \frac{x^2}{2}$$

where x is expressed in radians. These two approximations are valid to at least six decimal places for $|x| \le 0.01$, and to at least three decimals for $|x| \le 0.1$. They are useful for checking results or for estimating the value of the sine and cosine functions if a calculator or a set of trig tables is not available.

Example 1.35. Estimate the value of sin 3° and cos 3°.
We must first express the given argument (3°) in terms of radians:

$$x = \frac{3}{57.3} \approx 0.052$$

We can now make use of the small-argument approximations, which yield

$$\sin x \approx 0.052 \qquad \cos x \approx 1 - \frac{(0.052)^2}{2} \approx 0.9986$$

More precise values, obtained with an electronic calculator, are

$$\sin 3° = 0.052\,335\,96 \qquad \cos 3° = 0.998\,629\,54$$

Thus we see that the estimated values compare favorably with the more exact results.

The remaining four trigonometric functions—tangent, cotangent, secant, and cosecant—can easily be obtained from the sine and/or cosine through the following relationships:

$$\tan x = \frac{\sin x}{\cos x} \qquad \sec x = \frac{1}{\cos x}$$

$$\cot x = \frac{\cos x}{\sin x} \qquad \csc x = \frac{1}{\sin x}$$

Many electronic calculators allow the tangent (and its reciprocal, the cotangent) to be calculated directly, along with the sine and the cosine.

Example 1.36. Evaluate the sine, cosine, tangent, cotangent, secant, and cosecant of 0.75 radians.
Using an electronic calculator, we obtain

$$\sin 0.75 = 0.681\,638\,76 \qquad \cos 0.75 = 0.731\,688\,87$$

The remaining four functions can now be evaluated using the expressions presented above. Thus

$$\tan x = \frac{0.681\,638\,76}{0.731\,688\,87} = 0.931\,596\,46 \qquad \sec x = \frac{1}{0.731\,688\,87} = 1.366\,701\,1$$

$$\cot x = \frac{1}{0.931\,596\,46} = 1.073\,426\,2 \qquad \csc x = \frac{1}{0.681\,638\,76} = 1.467\,052\,7$$

Table 1-1.   Important Characteristics of Algebraic and Trigonometric Functions

Function	General Behavior	Key Values	Small-Argument Approximation
e^x	Increases rapidly as x increases ($x \geq 0$)	$e^0 = 1$ $\lim_{x \to \infty} e^x = \infty$	$e^x \approx 1 + x$
e^{-x}	Decays (approaches zero) as x increases ($x \geq 0$)	$e^0 = 1$ $\lim_{x \to \infty} e^{-x} = 0$	$e^{-x} \approx 1 - x$
$\log x$	Increases as x increases ($x > 0$) Assumes negative values for fractional values of x ($0 < x < 1$) Approaches $-\infty$ as x approaches 0 Not defined for $x < 0$ $\log x = (0.434\,294\,48)\ln x$	$\log 1 = 0$ $\lim_{x \to \infty} \log x = \infty$ $\lim_{x \to 0} \log x = -\infty$	
$\ln x$	Increases as x increases ($x > 0$) Assumes negative values for fractional values of x ($0 < x < 1$) Approaches $-\infty$ as x approaches 0 Not defined for $x < 0$ $\ln x = (2.302\,585\,1)\log x$	$\ln 1 = 0$ $\lim_{x \to \infty} \ln x = \infty$ $\lim_{x \to 0} \ln x = -\infty$	$\ln x \approx x - 1$ (x close to 1)
$\sin x$	Periodic with period 2π (radians) Antisymmetric, i.e. $\sin(-x) = -\sin x$ $\sin x = \cos\left(\frac{\pi}{2} - x\right)$	$\sin 0 = 0$ $\sin(\pi/2) = 1$ $\sin \pi = 0$ $\sin(3\pi/2) = -1$ $\sin 2\pi = 0$	$\sin x \approx x$
$\cos x$	Periodic with period 2π (radians) Symmetric, i.e. $\cos(-x) = \cos x$ $\cos x = \sin\left(\frac{\pi}{2} - x\right)$	$\cos 0 = 1$ $\cos(\pi/2) = 0$ $\cos \pi = -1$ $\cos(3\pi/2) = 0$ $\cos 2\pi = 1$	$\cos x \approx 1 - \frac{x^2}{2}$

1.12   THE ELECTRONIC CALCULATOR

There are many different kinds of electronic pocket calculators currently on the market. They vary considerably in price and complexity, ranging from the simple, inexpensive "four-function" calculators (add, subtract, multiply and divide) at one end of the spectrum, to the highly sophisticated, programmable calculators at the other. Between the two extremes are the so-called "electronic slide rule" or "scientific" calculators (Fig. 1-5). This type of calculator is especially well suited for the serious science or engineering student.

Virtually all scientific calculators include a provision for a floating decimal point and scientific notation. A typical scientific calculator will be able to display a number as large as $9.999\,99 \times 10^{99}$ or as small as $0.000\,01 \times 10^{-99}$. The use of scientific notation is optional, however, on most calculators.

Fig. 1-5. *Courtesy of Hewlett-Packard.*

Example 1.37. The quantity 0.006 25 can be entered into a scientific pocket calculator and displayed either in conventional form, as 0.006 25, or in scientific notation, as 6.25−03.

Most calculators have either an 8-digit or a 10-digit display. Thus numerical quantities can be represented to an accuracy of eight or ten significant figures. When a number is expressed in scientific notation, however, the last three digits may be reserved for the exponent, and so the mantissa may be reduced to a 5- or 7-digit number. (Some calculators employ an 8- or 10-digit mantissa *plus* a signed, 2-digit exponent.) The mantissa is usually expressed as a number whose magnitude lies between 1 and 9.999... .

Example 1.38. Shown below are several numbers as they would appear in conventional notation and in scientific notation on a typical 8-digit calculator.

Conventional Notation	Scientific Notation
1000.	1. 03
0.001	1.−03
123456.78	1.2346 05
−123456.78	−1.2346 05
0.00123456	1.2346−03
−0.00123456	−1.2346−03

When converting a number from conventional to scientific notation, most calculators will automatically round the mantissa if necessary, as illustrated above.

Operating Systems

A calculator's *operating system* governs the order in which information is keyed into the calculator and the order in which the calculations are carried out. Virtually all scientific calculators make use of either the Algebraic Operating System (AOS) or the Reverse Polish Notation (RPN) system.

Algebraic Operating System

The AOS allows information to be keyed into the calculator in the same order as a problem would be written down on paper.

Example 1.39. Carry out the calculation $5+3-6 = ?$, using a scientific, AOS electronic calculator.

The AOS keying sequence for carrying out this calculation is indicated below. (Each square containing a symbol represents a key on the calculator keyboard.)

Key	Display
5	5.
+	5.
3	3.
−	8.
6	6.
=	2.

The desired answer, 2., is shown in the display after the $=$ key is pressed.

AOS calculators make use of a natural hierarchy of calculations in which the operations are carried out in the following order:

1. evaluation of functions
2. exponentiation
3. multiplication and division
4. addition and subtraction

This natural hierarchy enhances the ability of the calculator to carry out the operations in the same order as they are written down.

Example 1.40. Evaluate the formula $y = 100 + 1200e^{-0.1}$, using an 8-digit, AOS electronic calculator.

The AOS keying sequence is indicated below.

Key	Display		Key	Display
1	1.		0	1200.
0	10.		×	1200.
0	100.		.	0.
+	100.		1	0.1
1	1.		+/−	−0.1
2	12.		e^x	0.90483742
0	120.		=	1185.8049

Thus the final answer, to 8 significant figures, is $y = 1185.8049$. In the calculation the exponential function is evaluated first, then this value is multiplied by 1200, and finally, 100 is added to the product.

Note the use of the $+/−$ key to change the sign of the number in the display.

Most AOS calculators include right- and left-hand parenthesis keys.   The use of parentheses permits the natural hierarchy to be overridden, thus allowing a long chain of operations to be carried out in any desired order.

Example 1.41.   Carry out the following calculation using a scientific, AOS electronic calculator:

$$\frac{2\times(8-3)\times\ln 4.8}{\sqrt{(1.63)^2-(0.79)^2}} = ?$$

An efficient AOS keying sequence is indicated below.

Key	Display		Key	Display
2	2.		1	1.
×	2.		.	1.
(	2.		6	1.6
8	8.		3	1.63
−	8.		x^2	2.6569
3	3.		−	2.6569
)	5.		.	0.
×	10.		7	0.7
4	4.		9	0.79
.	4.		x^2	0.6241
8	8.		)	2.0328
ln x	1.5686159		$\sqrt{x}$	1.425763
÷	15.686159		=	11.001941
(	15.686159			

Thus the desired answer, 11.001 941, is shown in the display after the ☐ = key has been pressed.

The keying sequences shown in Examples 1.39 through 1.41 are intended to be *representative*; there may actually be slight variations in the keying sequences between one AOS calculator and another.

Reverse Polish Notation

Calculators that make use of the RPN system include a "stack" of four registers (see Fig. 1-6). Each register can store one number.   When a number is first entered in the calculator it is placed at the bottom of the stack (in the X-register).   If a number had previously been in the X-register, it will be moved up to the Y-register to make room for the new number.   Similarly, a number originally in the Y-register will be moved up to the Z-register, and a number originally in the Z-register will move to the T-register (the top of the stack), replacing whatever number may originally have been there.   The contents of the X-register will always appear in the display.

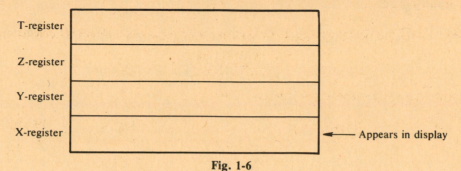

Fig. 1-6

Example 1.42.   The numbers 1, 3, 5, and 7 have been entered into an RPN calculator, in that order.   The number 9 is then entered.   The contents of the stack, before and after keying in the last number, are shown below.

	Before			After	
T	1		T	3	
Z	3		Z	5	
Y	5		Y	7	
X	7	←—Displayed	X	9	←—Displayed

The basic idea in Reverse Polish Notation is that the numbers are entered first, followed by the operation that is to be performed on them.   Arithmetic operations will always be carried out on the numbers in the X- and Y-registers.   The two operands themselves will be lost, but the result of the operation will be stored in the X-register (and therefore displayed).

Example 1.43.   Carry out the following calculation, using a scientific, RPN electronic calculator:

$$5+3-6 = ?$$

The RPN keying sequence is indicated below.   (The vertical arrow indicates the key used to enter a number into the stack.)

Key	Display (X-Register)
5	5.
↑	5.
3	3.
+	8.
6	6.
−	2.

Thus the desired answer, 2., appears in the display (the X-register) at the end of the calculation.   (Compare with Example 1.39.)

In order to evaluate a function with an RPN calculator, the argument must first be placed in the X-register. When the desired function key is pressed, the value of the function will be stored in the X-register, replacing the argument.

Example 1.44.   Evaluate the formula $y = \log_{10}(5+3-6)$ using an RPN calculator.

Key	Display (X-Register)
5	5.
↑	5.
3	3.
+	8.
6	6.
−	2.
log x	0.30103

Thus the desired answer is $y = 0.301\ 03$.

The order in which calculations are carried out with an RPN calculator is always determined by the order in which information is entered and stored within the stack.   Therefore a fixed hierarchy of operations is not required, nor is there a need for parentheses.

Example 1.45.   Evaluate the formula $y = 100 + 1200e^{-0.1}$ using an 8-digit, RPN calculator.

Key	Display (X-Register)	Key	Display (X-Register)
1	1.	↑	1200.
0	10.	·	0.
0	100.	1	0.1
↑	100.	+/−	−0.1
1	1.	e^x	0.90483742
2	12.	×	1085.8049
0	120.	+	1185.8049
0	1200.		

Thus the final answer is $y = 1185.8049$.   (Compare with Example 1.40.)

Example 1.46.   Carry out the following calculation using a scientific, RPN electronic calculator:

$$\frac{2\times(8-3)\times \ln 4.8}{\sqrt{(1.63)^2-(0.79)^2}} = \ ?$$

An efficient RPN keying sequence is indicated below.

Key	Display (X-Register)	Key	Display (X-Register)
2	2.	.	1.
↑	2.	6	1.6
8	8.	3	1.63
↑	8.	x^2	2.6569
3	3.	↑	2.6569
−	5.	.	0.
×	10.	7	0.7
4	4.	9	0.79
.	4.	x^2	0.6241
8	4.8	−	2.0328
ln x	1.5686159	$\sqrt{x}$	1.4257630
×	15.686159	÷	11.001941
1	1.		

Thus the desired answer, 11.001 941, is stored in the X-register and displayed after the ÷ key has been pressed.   (Compare with Example 1.41.)

It is difficult to say which of the two operating systems is better suited to the beginning student. Some persons prefer the simplicity of AOS while others favor the conciseness of RPN. The most important consideration, when comparing the two systems, is as to which one allows the user to visualize the proper order of entering information and carrying out operations most easily.

Programmable Pocket Calculators

These devices can be very useful for practicing scientists and engineers, particularly if a library of frequently used programs is available.   On the other hand, programmable calculators are expensive, and they are difficult to program.   (In fact, it is considerably easier to program a large computer, using a general-purpose algebraic language such as BASIC or FORTRAN.   We will say more about this in Chapter 7.)   Programmable calculators are therefore not recommended for the beginning science or engineering student.

Solved Problems

1.1. Write each of the following numbers in scientific notation.

(a)	0.85	(c)	−2 607 566	(e)	0.007 171
(b)	923.049	(d)	−50.63	(f)	−0.000 000 043 9

(a)	8.5×10^{-1}	(c)	$-2.607\,566 \times 10^6$	(e)	7.171×10^{-3}
(b)	$9.230\,49 \times 10^2$	(d)	-5.063×10^1	(f)	-4.39×10^{-8}

1.2. Write each of the numbers in Problem 1.1 in normalized floating point notation.

(a)	0.85E+0	(c)	−0.260 756 6E+7	(e)	0.7171E−2
(b)	0.923 049E+3	(d)	−0.5063E+2	(f)	−0.439E−7

1.3. Write each of the following numbers in conventional notation.

(a)	7.671×10^{-4}	(c)	0.5482E+2	(e)	$-5.052\,918 \times 10^4$
(b)	-2.314×10^3	(d)	0.9745E+6	(f)	−0.4070E−2

(a)	0.000 767 1	(c)	54.82	(e)	−50 529.18
(b)	−2314	(d)	974 500	(f)	−0.004 070

1.4. How many significant figures are present in each of the following numbers?

(a)	0.85	(c)	−2 607 566	(e)	0.007 171
(b)	923.049	(d)	5.0000	(f)	0.0100

(a)	two	(c)	seven	(e)	four
(b)	six	(d)	five	(f)	three

1.5. Round each of the following numbers to three significant figures.

(a)	372.1	(c)	−0.3415	(e)	72.749
(b)	5.719	(d)	−2 607 566	(f)	0.408 52

(a)	372.	(c)	−0.342	(e)	72.7
(b)	5.72	(d)	−2 610 000	(f)	0.409

1.6. Truncate each of the numbers given in Problem 1.5 to three significant figures.

(a)	372.	(c)	−0.341	(e)	72.7
(b)	5.71	(d)	−2 600 000	(f)	0.408

1.7. Carry out each of the calculations indicated below.

(a)	$(5.264 \times 10^3) + (6.18 \times 10^2)$	(d)	$(2.407 \times 10^7) \div (8.026 \times 10^5)$
(b)	$(2.886 \times 10^{-3}) - (0.465 \times 10^{-2})$	(e)	$(6.943 \times 10^4)^{0.5}$
(c)	$(7.23 \times 10^{-4}) \times (3.033 \times 10^2)$		

(a)
$$
\begin{array}{r}
5.264 \times 10^3 \\
+\ 0.618 \times 10^3 \\
\hline
5.882 \times 10^3
\end{array}
$$

The correct answer, to three significant figures, is 5.88×10^3.

(b)
$$2.8860 \times 10^{-3}$$
$$- \; 4.65 \;\; \times 10^{-3}$$
$$\overline{- \; 1.7640 \times 10^{-3}}$$

The correct answer, to three significant figures, is -1.76×10^{-3}.

(c)
$$(7.23 \times 3.033) \times 10^{-4+2} = 21.9 \times 10^{-2} = 0.219$$

(to three significant figures).

(d)
$$\frac{2.407}{8.026} \times 10^{7-5} = 0.2999 \times 10^2 = 29.99$$

(to four significant figures).

(e)
$$(6.943)^{0.5} \times 10^{(0.5)(4)} = 2.635 \times 10^2 = 263.5$$

(to four significant figures).

1.8. Evaluate each of the following functions.

      (a) $\sin(-38.4°)$ (c) $e^{\ln 6}$ (e) $\ln 10$
      (b) $\sec(-38.4°)$ (d) $\tan 3.95 \,(\text{rad})$

(a)
$$\sin(-38.4°) = -\sin 38.4° = -0.621\,15$$

(b)
$$\sec(-38.4°) = \frac{1}{\cos(-38.4°)} = \frac{1}{\cos 38.4°} = \frac{1}{0.783\,69} = 1.276\,01$$

(c)
$$e^{\ln 6} = 6$$

(d)
$$\tan 3.95 = \frac{\sin 3.95}{\cos 3.95} = \frac{-0.723\,19}{-0.690\,65} = 1.047\,11$$

(e)
$$\ln 10 = 2.302\,59 \log 10 = (2.302\,59)(1) = 2.302\,59$$

(This function can also be evaluated directly, using a set of tables or a calculator.)

1.9. Evaluate each of the following functions using an appropriate small-argument approximation.

      (a) $e^{0.005}$ (c) $\log 1.005$ (e) $\cos 0.005 \,(\text{rad})$
      (b) $\ln 1.005$ (d) $\sin 0.005 \,(\text{rad})$ (f) $\tan 0.005 \,(\text{rad})$

(a)
$$e^{0.005} \approx 1 + 0.005 = 1.005$$

(b)
$$\ln 1.005 \approx 1.005 - 1 = 0.005$$

(c)
$$\log 1.005 = 0.434\,29 \ln 1.005 \approx (0.434\,29)(0.005) \approx 0.002$$

(d)
$$\sin 0.005 \approx 0.005$$

(e)
$$\cos 0.005 \approx 1 - \frac{(0.005)^2}{2} = 0.999\,987\,5$$

(f)
$$\tan 0.005 \approx \frac{0.005}{0.999\,987\,5} \approx 0.005$$

1.10. Obtain an approximate answer for each of the following problems, *without* the use of a calculator or detailed hand calculations.

    (a) $\dfrac{1.669 \times 10^8 - 5.041 \times 10^7}{4.973 + 9.966 \times 10^{-2}}$ (b) $\dfrac{\sqrt{48\,524 + 79\,456}}{598.8}$ (c) $e^{-4.75/87.2} \sin\left(\dfrac{4.75}{65.4}\right)$

(a)
$$\frac{1.7 \times 10^8 - 0.5 \times 10^8}{5.0 + 0.1} = \frac{1.2}{5.1} \times 10^8 \approx 0.2 \times 10^8$$

(b)
$$\sqrt{\frac{4.9 \times 10^4 + 7.9 \times 10^4}{6.0 \times 10^2}} = \sqrt{\frac{12.8}{6.0}} \times 10^1 \approx \sqrt{2.1} \times 10^1 \approx 1.45 \times 10^1 = 14.5$$

(c) $e^{-(4.8/8.7)\times 10^{-1}} \sin\left(\frac{4.8}{6.5}\times 10^{-1}\right) \approx e^{-0.055} \sin 0.075 \approx (1-0.055)(0.075) = (0.945)(0.75 \times 10^{-1}) \approx 0.07$

Supplementary Problems

Answers are provided at the end of the book.

1.11. Write each of the following numbers in scientific notation.

(a) -0.4214	(c) 82.31	(e) -414.03	(g) 0.100 000
(b) 0.000 005 029 6	(d) $-38\,651\,066$	(f) -0.1	(h) $-0.000\,891\,1$

1.12. Write each of the numbers in Problem 1.11 in normalized floating point notation.

1.13. Write each of the following quantities in conventional notation.

(a) 2.0×10^{-1}	(c) 0.6776E$-$3	(e) 1.748×10^{-4}	(g) -5.675×10^{-3}	(i) 0.2105E$+$8
(b) -2.0×10^{1}	(d) -0.6776E$+2$	(f) 6.451×10^{5}	(h) 0.1667E$-$6	(j) -0.1E-1

1.14. Determine the number of significant figures in each of the following quantities.

(a) 100	(c) -0.001	(e) 0.1E$+6$	(g) 2.408×10^{-4}	(i) 210 505
(b) 100.	(d) 1.010	(f) -0.100	(h) $-0.363\,21$E-12	(j) 0.999 999

1.15. Round each of the following numbers to three significant figures.

(a) 128.6	(c) 0.998 54	(e) 0.9999	(g) 1.0051	(i) $-0.113\,47$
(b) $-0.004\,832$	(d) 0.9995	(f) 1.005	(h) 42 581 712	(j) 0.800 52

1.16. Truncate each of the numbers given in Problem 1.15 to three significant figures.

1.17. Carry out each of the calculations indicated below.

(a) $(4.293 \times 10^{-4}) + (6.406 \times 10^{-5})$

(b) $(3.130 \times 10^3) - (2.882 \times 10^2)$

(c) $(9.234 \times 10^6) \times (5.705 \times 10^{-2})$

(d) $(1.176 \times 10^{-3}) \div (1.346 \times 10^2)$

(e) $(4.852 \times 10^7)^{1/3}$

(f) $\dfrac{5.412 \times 10^6 + 9.539 \times 10^4}{4.412 \times 10^5 - 8.593 \times 10^6}$

(g) $[(4.786 \times 10^{-2})(9.458 \times 10^{-1}) + (5.087 \times 10^{-1})(0.7079 \times 10^{-2})]^{-1/2}$

(h) $[(563.9)^2 + (69.7)^2 + (90.7)^2 + (200.8)^2]^{1/2}$

1.18. Obtain an approximate answer for each of the following problems, *without* the use of a calculator or detailed hand calculations.

(a) $[(6.854 \times 10^6)(8.268 \times 10^{-8}) + (9.257 \times 10^{-4})(5.168 \times 10^3)]^{1/2}$

(b) $\dfrac{8.941 \times 10^3 + 8.126 \times 10^4}{6.517 \times 10^5 - 6.448 \times 10^5}$

(c) $(2.54)^6$

(d) $(68.3)^{-0.8}$

1.19. Convert from degrees to radians: (a) $17.5°$, (b) $-200°$, (c) $360°$, (d) $500°$.

1.20. Convert from radians to degrees: (a) 0.1, (b) 3, (c) −0.05, (d) 8.

1.21. Evaluate each of the following *without* using a calculator or referring to the text.

(a) e^0 (d) $\ln 0$ (g) $\sin \pi$ (j) $\sin (\pi/2)$ (m) $\sin (3\pi/2)$

(b) $\log 1$ (e) $\sin 0$ (h) $\cos 0$ (k) $\cos (\pi/2)$ (n) $\cos 2\pi$

(c) e^{-1000} (f) $\cos (3\pi/2)$ (i) $\sin 2\pi$ (l) $\cos \pi$

1.22. What is the relationship between $\sin x$ and $\sin (-x)$?

1.23. What is the relationship between $\cos x$ and $\cos (-x)$?

1.24. What is the relationship between $\sin x$ and $\cos x$?

1.25. Evaluate each of the following functions using an appropriate small-argument approximation. (The arguments of the trigonometric functions are in radians.)

(a) $e^{-0.035}$ (c) $\log 1.035$ (e) $\cos (-0.035)$

(b) $\ln 1.035$ (d) $\sin (-0.035)$ (f) $\tan 0.035$

1.26. Carry out each of the following calculations as accurately as possible. (The arguments of the trigonometric functions are in radians.)

(a) $\sin 5.62 + \cos 5.62$ (c) $0.3 e^2 + 0.7 e^3 + 0.1 e^{-1}$

(b) $e^{-0.667} \sin 0.667$ (d) $5 - \ln \sqrt{(1.87)^2 + (2.05)^2}$ (e) $\dfrac{1}{3\sqrt{2\pi}} e^{-[(8.50-1.67)^2/18]}$

(f) $0.25 \cos \left(\ln \dfrac{\pi}{4} \right) + 0.75 \sin \left(\ln \dfrac{3\pi}{4} \right)$

1.27. Refer to Example 1.26. (a) What will be the current after 1 second? (b) What fraction of the initial current will remain after 5 seconds? (c) How long will it take for the current to reach 15 percent of its initial value? (d) Sketch a plot of current versus time.

1.28. Refer to Example 1.27. (a) What will be the concentration of B after 3 hours? (b) What will be the concentration of B after 10 hours? (c) At what time will the concentration of B be maximized? (d) What will be the maximum concentration of B? (e) Sketch a plot of concentration of B versus time.

1.29. Refer to Example 1.28. (a) What height of water corresponds to a discharge velocity of 0.04 m/s? (b) What will be the discharge velocity when the height of water in the tank is 0.10 m? (c) Sketch a plot of height of water versus discharge velocity.

1.30. Refer to Example 1.29. (a) Determine the temperature at the center of the rod 5 s after the rod is plunged into the oil bath. (b) What will be the temperature after 30 s? (c) What will be the temperature after a very long period of time? (d) How long will it take for the center of the rod to reach 300 °C? (e) Sketch a plot of rod temperature versus time.

1.31. The voltage drop across an electronic device can be calculated using the formula

$$v = 0.5 e^{-0.2t} \sin 0.1t$$

where v is the voltage, in millivolts, and t is the time, in seconds, after the actuating switch is closed. (a) Determine the voltage drop at the instant the activating switch is closed. (b) Determine the voltage drop after 10 s. (c) Determine the voltage drop after 1 min. (d) Estimate the maximum voltage drop, and determine when this will occur. (e) What will be the voltage drop after a very long period of time? (f) Sketch a plot of voltage drop versus time.

1.32. The keying sequences for two problems being solved on electronic calculators are shown below. For each problem, specify the type of calculator being used (AOS or RPN) and determine what problem is being solved.

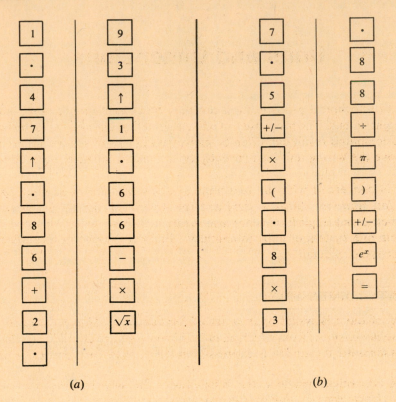

(a)   (b)

Chapter 2

Units and Dimensions

The solution to a technical problem must not only be correct numerically, but it must also be expressed in the right units. The proper use of units is sometimes confusing to the beginning student, however, since a computed result can often be expressed in any of several different units, and it is very easy to choose the wrong unit inadvertently or to make a mistake when converting from one unit to another.

This chapter is concerned with the two units systems that are most often used for technical calculations, the *international metric system* and the *American engineering system*. Some other commonly used metric and English units are also mentioned. An orderly method is then presented for converting from one system of units to another. Finally, the important subject of dimensional consistency is discussed in detail.

2.1 FUNDAMENTAL DIMENSIONS

Every system of units is based upon a set of *fundamental dimensions* (also called *reference dimensions* or *base dimensions*) from which all other dimensions are derived. Within a given system of units, each fundamental dimension is expressed in terms of its own particular unit.

Example 2.1. The international metric system is based upon the following seven fundamental dimensions: *length, mass, time, electric current, absolute (thermodynamic) temperature, luminous intensity*, and *amount of substance*. This last dimension, which was lacking in earlier versions of the system, may be ignored for most purposes.

In some units systems, force may be substituted for mass, or electric charge for electric current. The American engineering system includes both force and mass as fundamental dimensions.

Example 2.2. The American engineering system is based upon the following seven fundamental dimensions: *length, mass, time, force, electric charge, absolute (thermodynamic) temperature*, and *luminous intensity*. Moreover, most of these dimensions are expressed in different units than the corresponding dimensions in the international metric system (e.g. the metric unit of length is the *meter*, whereas the American engineering unit of length is the *foot*).

It is convenient to represent each fundamental dimension with its own *symbol*, enclosed in square brackets. These symbols, which are independent of any particular system of units, are listed in Table 2-1. It should be noted that no single system of units incorporates all the fundamental dimensions shown in Table 2-1.

Table 2-1

Fundamental Dimension	Symbol	Fundamental Dimension	Symbol
length	$[L]$	electric charge	$[Q]$
mass	$[M]$	electric current	$[A]$
time	$[T]$	temperature (absolute)	$[\theta]$
force	$[F]$	luminous intensity	$[I]$

2.2 DERIVED DIMENSIONS

Most of the dimensions within a given units system are defined in terms of some combination of two or more fundamental dimensions.   Such dimensions are called *derived dimensions*.   Symbolically, a derived dimension can be expressed as a product or quotient involving two or more fundamental symbols.

Example 2.3.   (*a*)   *Area* can be defined as the product of one length times another (i.e. length times width).   Symbolically, this can be expressed as $[L \cdot L]$, or simply $[L^2]$.   (*b*)   *Velocity* is defined as the distance traveled (in a particular direction) per unit time.   Symbolically, this can be expressed as $[L/T]$ (length divided by time).   Thus velocity is a derived dimension.

Table 2-2 shows the symbolic representations for a number of the more common derived dimensions.   Some of these derived dimensions can be expressed in different ways, depending on the particular choice of fundamental dimensions.   Multiple entries are shown in such cases.

Table 2-2

Derived Dimension	Symbol
area	$[L^2]$
volume	$[L^3]$
mass	$[FT^2/L]$
force	$[ML/T^2]$
velocity	$[L/T]$
acceleration	$[L/T^2]$
angular velocity	$[T^{-1}]$ (radians per unit time)
angular acceleration	$[T^{-2}]$ (radians per unit time per unit time)
momentum	$[ML/T]$ or $[FT]$
pressure	$[M/LT^2]$ or $[F/L^2]$
stress	$[M/LT^2]$ or $[F/L^2]$
energy (work)	$[ML^2/T^2]$ or $[FL]$
torque	$[ML^2/T^2]$ or $[FL]$
power	$[ML^2/T^3]$ or $[FL/T]$
density	$[M/L^3]$ or $[FT^2/L^4]$
viscosity	$[M/LT]$ or $[FT/L^2]$
kinematic viscosity	$[L^2/T]$
heat capacity	$[L^2/T\theta]$
thermal conductivity	$[ML/T^3\theta]$ or $[F/T\theta]$
frequency	$[T^{-1}]$ (cycles per unit time)
electric charge	$[AT]$
electric current	$[Q/T]$
electric potential (voltage)	$[ML^2/AT^3]$, $[ML^2/QT^2]$, $[FL/AT]$ or $[FL/Q]$
electric resistance	$[ML^2/A^2T^3]$, $[ML^2/Q^2T]$, $[FL/A^2T]$ or $[FLT/Q^2]$
electric capacitance	$[A^2T^4/ML^2]$, $[Q^2T^2/ML^2]$, $[A^2T^2/FL]$ or $[Q^2/FL]$
electric inductance	$[ML^2/A^2T^2]$, $[ML^2/Q^2]$, $[FL/A^2]$ or $[FLT^2/Q^2]$

Example 2.4.   *Work* is defined as the product of force and distance (in the direction of the force).   In a units system that includes force as a fundamental dimension, such as the American engineering system, work can be expressed simply as $[FL]$.   If force is a derived dimension, as in the international metric system, then it

must be expressed symbolically as $[ML/T^2]$ (i.e. mass times acceleration, where acceleration has the dimensions $[L/T^2]$). Therefore the symbolic representation for work becomes

$$[ML/T^2] \cdot [L] = [ML^2/T^2]$$

It is noteworthy that work has the same units as *energy*. Thus energy can be expressed symbolically as $[FL]$ or as $[ML^2/T^2]$, depending upon the particular system of units. Both of these forms are included in Table 2-2.

Example 2.5. *Electric potential* can be defined as work per unit of electric charge. Now, work can be expressed in terms of either force $[F]$ or mass $[M]$, and electric charge can either be taken as a fundamental dimension $[Q]$ or it can be expressed in terms of electric current $[A]$. This gives $2 \times 2 = 4$ ways of symbolizing electric potential, as shown in Table 2-2.

Within a given units system many derived dimensions will be expressed in terms of single units. It is important to remember that such units are always defined in terms of other units which represent fundamental dimensions.

Example 2.6. In the international metric system the unit of force is the *newton* (N). As shown in Example 2.4,

$$[F] = [ML/T^2]$$

The fundamental dimensions on the right have as units the *kilogram* (kg), the *meter* (m), and the *second* (s), respectively. Thus

$$1 \text{ N} = 1 \text{ kg} \cdot \text{m/s}^2$$

2.3 THE INTERNATIONAL METRIC SYSTEM (SI UNITS)

The *international metric system* or *SI*, as it is universally abbreviated, is a variation of the metric *mks* (*meter - kilogram - second*) system, which has been used in the natural sciences for many years. SI now serves as the standard for engineering calculations in most parts of the world. Within the English-speaking countries, however, the engineering communities have historically made use of an entirely different system of units. Efforts are now under way in these countries to replace all other units systems with SI.

Table 2-3 lists the SI base units (which represent fundamental dimensions) and Table 2-4 defines several of the more common derived units.

Table 2-3. SI Base Units

Fundamental Dimension	Base Unit
length $[L]$	meter (m)
mass $[M]$	kilogram (kg)
time $[T]$	second (s)
electric current $[A]$	ampere (A)
temperature $[\theta]$	kelvin (K)
luminous intensity $[I]$	candela (cd)
amount of substance	mole (mol)

Example 2.7. (*a*) The definitions in Table 2-4 involve not only base units but also already-defined derived units. Thus

$$1 \text{ J} = 1 \text{ N} \cdot \text{m} = 1 \text{ (kg} \cdot \text{m/s}^2) \cdot \text{m} = 1 \text{ kg} \cdot \text{m}^2/\text{s}^2$$

and this expression for the joule in terms of *base* units agrees with $[ML^2/T^2]$, the dimensions of energy. (*b*) The definition of the lumen involves one of the two SI *supplementary units*, the *steradian* (sr). This is a unit of solid angle such that at the center of a sphere, the solid angle subtended by the surface is 4π steradians. The other supplementary unit is the *radian* (rad), which measures plane angles.

Table 2-4.   SI Derived Units

Derived Dimension	Unit	Definition
force $[ML/T^2]$	newton (N)	kg $\cdot$ m/s^2
pressure $[M/LT^2]$	pascal (Pa)	N/m^2
energy $[ML^2/T^2]$	joule (J)	N $\cdot$ m
power $[ML^2/T^3]$	watt (W)	J/s
frequency $[T^{-1}]$	hertz (Hz)	s^{-1}
electric charge $[AT]$	coulomb (C)	A $\cdot$ s
electric potential $[ML^2/AT^3]$	volt (V)	W/A
electric resistance $[ML^2/A^2T^3]$	ohm (Ω)	V/A
electric capacitance $[A^2T^4/ML^2]$	farad (F)	A $\cdot$ s/V
electric inductance $[ML^2/A^2T^2]$	henry (H)	V $\cdot$ s/A
magnetic flux $[ML^2/AT^2]$	weber (Wb)	V $\cdot$ s
magnetic flux density $[M/AT^2]$	tesla (T)	Wb/m^2
luminous flux $[I]$	lumen (lm)	cd $\cdot$ sr
illumination $[I/L^2]$	lux (lx)	lm/m^2

Example 2.8   (*a*)   A mass of 4 kg is accelerated at 3 m/s^2.   The corresponding force is

$$(4 \text{ kg})(3 \text{ m/s}^2) = 12 \text{ kg} \cdot \text{m/s}^2 = 12 \text{ N}$$

(*b*)   The electric charge required to produce a current of 5 amperes for 3 seconds is

$$(5 \text{ A})(3 \text{ s}) = 15 \text{ A} \cdot \text{s} = 15 \text{ C}$$

Other Metric Units

There are two commonly used metric units that are not included in Table 2-4.   The *degree Celsius* (°C) is identical in magnitude to the kelvin (K).   It is used to measure temperatures or temperature intervals on the Celsius scale, which differs from the Kelvin scale only in the choice of zero-point.   The two scales are related by

$$\theta_C = \theta_K - 273.15$$

Water, at sea level, will freeze at 0 °C and boil at 100 °C.

Example 2.9.   What is the normal boiling point of water, in kelvins?
Water normally boils at 100 °C (at sea level).   Hence

$$\theta_K = \theta_C + 273.15 = 100 + 273.15 = 373.15$$

Therefore the boiling point of water is approximately 373 K.

The *kilocalorie* is defined as the heat (energy) required to raise the temperature of 1 kg of water 1 K (or 1 °C).   One kcal is equivalent to 4184 joules.

Example 2.10.   How much energy is required to raise the temperature of 5 kg of water from 300 K to 350 K?
The amount of energy required is proportional to the mass of the water and to the increase in its temperature:

$$\text{energy} = (\text{mass}) \times (\text{specific heat}) \times (\text{temperature increase})$$
$$= (5 \text{ kg})(1 \text{ kcal/kg} \cdot \text{K})(350 \text{ K} - 300 \text{ K}) = 250 \text{ kcal}$$

Converting this result to joules, we have

$$\text{energy} = (250 \text{ kcal})(4184 \text{ J/kcal}) = 1.046 \times 10^6 \text{ J}$$

Thus, about one million joules is required to raise the temperature of 5 kg of water 50 K.

A more extensive discussion of how to convert units will be given in Section 2.5.

SI Prefixes

SI also makes use of unit prefixes, which represent decimal multipliers of the units to which they are attached.   Table 2-5 contains the more commonly used SI prefixes.

Table 2-5

Prefix	Decimal Multiplier	Symbol
giga	10^9	G
mega	10^6	M
kilo	10^3	k
deci*	10^{-1}	d
centi*	10^{-2}	c
milli	10^{-3}	m
micro	10^{-6}	μ
nano	10^{-9}	n
pico	10^{-12}	p

*Permitted, but not preferred.

Example 2.11.   (a)   One *megawatt* (MW) is equal to 1 000 000 W, since the prefix *mega* represents the multiplier 10^6.   (b)   One *centimeter* (cm) is equal to 0.01 m, since the prefix *centi* represents the multiplier 10^{-2}. Hence there are 100 centimeters in one meter.   (c)   One *microsecond* (μs) is equal to 10^{-6} s, since the prefix *micro* represents the multiplier 10^{-6}.   Thus we see that one second contains 1 000 000 microseconds.

2.4 THE AMERICAN ENGINEERING SYSTEM

The *American engineering system* is a variation of the *English fps* (*foot-pound-second*) system of units.   Several more decades may be required before this system is replaced by SI in the English-speaking countries.   We will therefore consider the American engineering system in some detail.

Table 2-6 indicates the base units that represent the various fundamental dimensions of the American engineering system.   A few of the more common derived dimensions are shown in Table 2-7.   Certain other derived dimensions, such as electric potential, electric resistance, magnetic flux and luminous flux, employ the same units as in SI (i.e. volts, ohms, webers and lumens).   The basic time unit, the second, and the derived *hour* are identical in SI and the American engineering system.   However, these units are abbreviated in SI as s and h, whereas in the American system the usual abbreviations are sec and hr.

Table 2-6.   American Engineering Base Units

Fundamental Dimension	Base Unit
length [L]	foot (ft)
mass [M]	pound (lb_m)
time [T]	second (sec)
force [F]	pound (lb_f)
electric charge [Q]	coulomb (C)
temperature [θ]	degree Rankine (°R)
luminous intensity [I]	candle (cd)

Table 2-7. American Engineering Derived Units

Derived Dimension	Unit	Definition
density $[M/L^3]$	—	lb_m/ft^3
pressure $[F/L^2]$	—	lb_f/ft^2
energy $[FL]$	—	$ft \cdot lb_f$
power $[FL/T]$	—	$ft \cdot lb_f/sec$
electric current $[Q/T]$	ampere (A)	C/sec

Mass and force are both considered to be fundamental dimensions in the American engineering system. To further complicate matters, the base unit for *each* of these dimensions is the *pound*. Thus there are units of *pounds mass* (lb_m) and *pounds force* (lb_f) in the American engineering system. The relationship between them is that a mass of one pound (1 lb_m) will experience a gravitational force of one pound (i.e. will weigh 1 lb_f) on the earth, at sea level and a latitude of 45°.

Example 2.12. Determine the weight of   (*a*) a 5 lb_m mass and   (*b*)   a 5 kg mass, at sea level and a latitude of 45°.

(*a*) Since 1 lb_m will weigh 1 lb_f under these conditions, 5 lb_m will weigh 5 lb_f.

(*b*) In SI, force, a derived dimension, is defined by Newton's second law of motion in the familiar form $F = ma$. The gravitational acceleration at sea level and a latitude of 45° is 9.807 m/s². Hence the weight of the 5 kg mass will be

$$(5 \text{ kg})(9.807 \text{ m/s}^2) = 49.035 \text{ N}$$

Notice that in (*a*) the gravitational constant (32.174 ft/sec²) was not required when converting pounds mass to pounds force (weight). The metric gravitational constant (9.807 m/s²) *is* required, however, when converting kilograms to newtons.

If, in the American engineering system, one tries to use Newton's second law in the form $F = ma$, the force comes out in *poundals*, a nonbasic unit, when the mass is expressed in lb_m. That is to say,

$$1 \text{ poundal} = 1 \text{ lb}_m \cdot ft/sec^2$$

In order to convert from poundals to lb_f, we must divide by the *gravitational constant,* which is $g_c = 32.174$ poundals/lb_f = 32.174 $lb_m \cdot ft/lb_f \cdot sec^2$. (This value is numerically equivalent to the gravitational acceleration on earth, at sea level and a latitude of 45°.) Thus, in the American engineering system, the correct form of Newton's second law is

$$F = \frac{ma}{g_c}$$

Example 2.13. Determine the weight of a 5 lb_m mass on the surface of the moon, where the gravitational acceleration is approximately 1/6 the value on earth.
   Newton's second law gives

$$F = \frac{ma}{g_c} = \frac{(5 \text{ lb}_m)(\frac{1}{6} \times 32.174 \text{ ft/sec}^2)}{32.174 \text{ lb}_m \cdot ft/lb_f \cdot sec^2} = \frac{5}{6} \text{ lb}_f$$

Notice that the units $lb_m \cdot ft/sec^2$ (or poundals) cancel, leaving only the desired units, lb_f. Also, the numerical value 32.174 cancels. Thus the correct result can be obtained simply by writing

$$F = (5)(\tfrac{1}{6}) = \frac{5}{6} \text{ lb}_f$$

even though the units conversions are not shown explicitly.

Use of the American engineering system is straightforward where conversions between mass and force are not required.

Example 2.14. How much work is done when a force of 25 lb_f acts through a distance of 40 ft?
Work is the product of force and distance, and so

$$\text{work} = (25 \text{ lb}_f)(40 \text{ ft}) = 1000 \text{ ft} \cdot \text{lb}_f$$

Example 2.15. Determine the electric current corresponding to a uniform flow of 10 C over a period of 5 sec.
Electric current is defined as the amount of electric charge flowing past a fixed point per unit time. Thus

$$\text{current} = \frac{10 \text{ C}}{5 \text{ sec}} = 2 \text{ C/sec} = 2 \text{ A}$$

Other English Units

There are many other English units used in technical calculations which, while not expressly included in the American engineering system, can be given in terms of the American engineering base units. For example, volume can be expressed in *gallons*, pressure in *atmospheres*, energy in *Btu*, etc. Table 2-8 lists several of the more commonly used English units of this type.

Table 2-8. Commonly Used English Units

Dimension	Unit	Definition
length [L]	inch (in)	12 in = 1 ft
	yard (yd)	1 yd = 3 ft
	mile (mi)	1 mi = 5280 ft
area [L^2]	acre	1 acre = 43 560 ft²
volume [L^3]	gallon (gal)	7.480 52 gal = 1 ft³
mass [M]	ounce (oz)	16 oz = 1 lb_m
	ton	1 ton = 2000 lb_m
time [T]	minute (min)	1 min = 60 sec
	hour (hr)	1 hr = 60 min = 3600 sec
	day	1 day = 24 hr = 86 400 sec
temperature [θ]	degree Fahrenheit (°F)	1 °F = 1 °R
		$\theta_F = \theta_R - 459.67$
pressure [F/L^2]	lb_f/in² (psi)	1 psi = 144 lb_f/ft²
	atmosphere (atm)	1 atm= 14.696 psi
		= 2116.224 lb_f/ft²
power [FL/T]	horsepower (hp)	1 hp = 550 ft $\cdot lb_f$/sec
energy [FL]	British thermal unit (Btu)	1 Btu = 777.65 ft $\cdot lb_f$
	horsepower-hour (hp $\cdot$ hr)	1 hp $\cdot$ hr = 1.98×10^6 ft $\cdot lb_f$
	kilowatt-hour (kw $\cdot$ hr)	1 kW $\cdot$ hr = $2.655\ 22 \times 10^6$ ft $\cdot lb_f$

Example 2.16. How many gallons of water can be contained in a cylinder having a radius (r) of 2 ft and a height (h) of 5 ft?
The volume of the cylinder is given by

$$V = \pi r^2 h = (3.141\ 59)(2)^2(5) = 62.831\ 85 \text{ ft}^3$$

From Table 2-8, 7.480 52 gal = 1 ft³. Hence

$$V = (62.831\ 85 \text{ ft}^3)(7.480\ 52 \text{ gal/ft}^3) = 470.015 \text{ gal}$$

We conclude that the cylinder will hold about 470 gal of water.

Example 2.17.   How much energy is required to raise the temperature of 5 lb_m of water from 300 °R to 350 °R?

The *British thermal unit* (Btu) is a unit of energy which is defined as the heat required to raise 1 lb_m of water 1 °R (or 1 °F).   This is analogous to the definition of the kilocalorie given in Section 2.3.   Then as in Example 2.10,

$$\text{energy} = (5 \ lb_m)(1 \ \text{Btu}/lb_m \cdot °R)(350 \ °R - 300 \ °R) = 250 \ \text{Btu}$$

Since   1 Btu = 777.65 ft · lb_f (from Table 2-8), we obtain

$$\text{energy} = (250 \ \text{Btu})(777.65 \ \text{ft} \cdot lb_f/\text{Btu}) = 1.94 \times 10^5 \ \text{ft} \cdot lb_f$$

Thus, about 194 000 ft · lb_f is required to raise the temperature of 5 lb_m of water 50 °R.   This is equivalent to about 263 000 joules (compare with Example 2.10).

2.5   UNITS CONVERSIONS

Frequently the beginner will attempt to convert units using intuitive, "seat-of-the-pants" methods, resulting in answers that are often incorrect.   This section presents an orderly method for converting units, based upon the use of *units equivalences*.   A collection of units equivalences (including some of the factors presented in Tables 2-4 and 2-8) is given in Appendix A.

Example 2.18.   How many square feet are equivalent to 1 square kilometer?

From the first entry in Appendix A, we see that

$$43 \ 560 \ \text{ft}^2 = 4.046 \ 856 \times 10^{-3} \ \text{km}^2$$

If we divide through by 4.046 856 × 10⁻³, we obtain

$$\frac{4.356 \times 10^4}{4.046 \ 856 \times 10^{-3}} \ \text{ft}^2 = 1 \ \text{km}^2$$

or

$$1.076 \ 391 \times 10^7 \ \text{ft}^2 = 1 \ \text{km}^2$$

Hence 1 square kilometer is equivalent to 10 763 910 square feet.

Another way to solve this problem is to make use of the information listed in Appendix A under the dimension *length*.   Thus   1 km = 1000 m   and   1 m = 3.280 84 ft.   If we multiply the second equivalence by 1000, we obtain

$$1 \ \text{km} = 1000 \ \text{m} = 3.280 \ 84 \times 10^3 \ \text{ft}$$

Squaring both sides,

$$1 \ \text{km}^2 = 10.763 \ 91 \times 10^6 \ \text{ft}^2 = 1.076 \ 391 \times 10^7 \ \text{ft}^2$$

which is the same result obtained earlier.

It is often desirable to write a units equivalence as a fraction, called an *equivalence factor* or a *conversion factor*.   A conversion factor has the value unity in multiplications.

Example 2.19.   Again referring to the first entry in Appendix A, we see that

$$1 \ \text{acre} = 43 \ 560 \ \text{ft}^2 = 4.046 \ 856 \times 10^{-3} \ \text{km}^2$$

We can form the following conversion factors from this information:

$$\frac{1 \ \text{acre}}{43 \ 560 \ \text{ft}^2} \qquad\qquad \frac{43 \ 560 \ \text{ft}^2}{1 \ \text{acre}}$$

$$\frac{1 \ \text{acre}}{4.046 \ 856 \times 10^{-3} \ \text{km}^2} \qquad\qquad \frac{4.046 \ 856 \times 10^{-3} \ \text{km}^2}{1 \ \text{acre}}$$

$$\frac{43 \ 560 \ \text{ft}^2}{4.046 \ 856 \times 10^{-3} \ \text{km}^2} \qquad\qquad \frac{4.046 \ 856 \times 10^{-3} \ \text{km}^2}{43 \ 560 \ \text{ft}^2}$$

To carry out a units conversion, the given quantity is multiplied by one or more conversion factors. The final product will be expressed in terms of the desired units provided the conversion factors have been chosen properly. These factors must be selected in such a manner that the *original* (*given*) units *cancel* one another, leaving only the new (desired) units.

Example 2.20. How many acres are equivalent to 15 square kilometers?

This problem can easily be solved by use of the appropriate conversion factor from Example 2.19:

$$(15 \ \text{km}^2)\left(\frac{1 \ \text{acre}}{4.046\,856 \times 10^{-3} \ \text{km}^2}\right) = 3.706\,581 \times 10^3 \ \text{acres}$$

Thus 15 square kilometers is equivalent to about 3707 acres.

Notice that the conversion factor was chosen so that the given units (km^2) would appear in both the numerator and the denominator, thus canceling each other. This leaves only the units acres, as desired.

Example 2.21. How many dynes are equivalent to a force of 50 pounds?

From the *force* entries in Appendix A, we obtain

$$(50 \ \text{lb}_f)\left(\frac{4.448\,22 \ \text{N}}{1 \ \text{lb}_f}\right)\left(\frac{1 \ \text{dyne}}{10^{-5} \ \text{N}}\right) = 2.224\,11 \times 10^7 \ \text{dynes}$$

Alternatively, we can write

$$(50 \ \text{lb}_f)\left(\frac{1 \ \text{N}}{0.224\,809 \ \text{lb}_f}\right)\left(\frac{1 \ \text{dyne}}{10^{-5} \ \text{N}}\right) = 2.224\,11 \times 10^7 \ \text{dynes}$$

Example 2.22. Convert a power output of 10 megawatts to an equivalent number of Btu per hour.

Referring to the *power* entries in Appendix A, we can write

$$(10 \ \text{MW})\left(\frac{10^6 \ \text{W}}{1 \ \text{MW}}\right)\left(\frac{1 \ \text{Btu/hr}}{0.292\,875 \ \text{W}}\right) = 3.414\,426 \times 10^7 \ \text{Btu/hr}$$

Thus 10 megawatts is equivalent to about 34 million Btu per hour.

Example 2.23. The heat capacity of a substance is 0.285 Btu/$\text{lb}_m \cdot \,^\circ\text{F}$. Express the heat capacity in terms of J/kg $\cdot$ K.

Appendix A contains the required units equivalence listed under *heat capacity*. Thus,

$$(0.285 \ \text{Btu/lb}_m \cdot \,^\circ\text{F})\left(\frac{1 \ \text{J/kg} \cdot \text{K}}{2.390\,06 \times 10^{-4} \ \text{Btu/lb}_m \cdot \,^\circ\text{F}}\right) = 1192.44 \ \text{J/kg} \cdot \text{K}$$

The above calculation was very straightforward, since we had a single conversion factor that allowed us to convert directly from the given units to the desired units. If such multi-unit conversion factors are not available, the calculation may be carried out using several single-unit conversion factors, as follows:

$$(0.285 \ \text{Btu/lb}_m \cdot \,^\circ\text{F})\left(\frac{1054.35 \ \text{J}}{1 \ \text{Btu}}\right)\left(\frac{2.204\,62 \ \text{lb}_m}{1 \ \text{kg}}\right)\left(\frac{1.8 \,^\circ\text{F}}{1 \ \text{K}}\right) = 1192.44 \ \text{J/kg} \cdot \text{K}$$

All the single-unit conversion factors appearing in this calculation were obtained from Appendix A, though they could have been obtained from many other sources. A particularly complete tabulation of conversion factors can be found in recent editions of the *Handbook of Chemistry and Physics*, published annually by the Chemical Rubber Co., Cleveland, Ohio.

2.6 DIMENSIONAL CONSISTENCY

An equation cannot be correct unless it is *dimensionally consistent*. This means that both sides of the equation must be expressed in terms of the same dimensions.

Example 2.24. If an object having mass m is raised to a height h above a horizontal surface and then released, the velocity upon impact with the horizontal surface can be calculated as $v = \sqrt{2gh}$, where g is the gravitational acceleration. Determine whether or not this equation is dimensionally consistent.

Since velocity has the dimensions $[L/T]$ (see Table 2-2), the right-hand side of the equation must also have the dimensions $[L/T]$. We know (Table 2-2) that gravitational acceleration has the dimensions $[L/T^2]$ and that height has the dimension $[L]$. The constant 2 is dimensionless. Therefore $\sqrt{2gh}$ has the dimensions

$$\sqrt{[L/T^2] \cdot [L]} = \sqrt{[L^2/T^2]} = [L/T]$$

as required. The equation is therefore dimensionally consistent.

The fact that an equation is dimensionally consistent does not guarantee that it is correct.

Example 2.25. Suppose that the equation in Example 2.24 had been written incorrectly as $v = \sqrt{gh}$. This equation is dimensionally consistent even though a factor of 2 has been omitted.

If an equation is not dimensionally consistent, then the equation must be incorrect. Thus a check for dimensional consistency can be very helpful when checking for errors.

Example 2.26. A student who is studying fluid mechanics has written the basic equation of hydrostatics as

$$\frac{\Delta p}{\rho} = -gh^2$$

where Δp is the change in fluid pressure corresponding to an increase in height h, ρ is the density of the fluid, and g is the gravitational acceleration. Is this equation dimensionally consistent?

From Table 2-2 we see that pressure has the dimensions $[M/LT^2]$, density has the dimensions $[M/L^3]$, and gravitational acceleration has the dimensions $[L/T^2]$. Thus the equation can be written dimensionally as

$$\frac{[M/LT^2]}{[M/L^3]} = [L/T^2] \cdot [L^2]$$

which can be simplified to read $[L^2/T^2] = [L^3/T^2]$. The equation is not dimensionally consistent, which means that it must be incorrect. (The correct form of the equation is $(\Delta p)/\rho = -gh$.)

Logarithmic, exponential, and trigonometric functions and their arguments are dimensionless (see Section 1.11).

Example 2.27. The current in a certain electrical circuit can be represented by the equation

$$y = ae^{-bt}$$

where y represents the current, t represents time, and a and b are positive constants. What are the proper dimensions for a and for b?

The exponent $-bt$ must be dimensionless. Since t has the dimension $[T]$, the constant b must have the dimension $[T^{-1}]$ for the dimensions to cancel when the product bt is formed.

The exponential function e^{-bt} will also be dimensionless. Thus the constant a must have the same dimension as y; namely $[A]$, which represents current.

We can check these results by expressing the original equation in dimensional form. Thus

$$[A] = [A]e^{-[1/T] \cdot [T]} = [A]$$

The equation is dimensionally consistent, since the dimensions cancel out of the exponent and we are left with the same dimension, $[A]$, on each side of the equation.

The concept of dimensional consistency can easily be applied to specific units. This allows us to check an equation for *units consistency*. It should be understood that this is a more demanding condition than dimensional consistency, since an equation that is dimensionally correct can still contain inconsistent units.

Example 2.28. The basic equation of hydrostatics can be written as $\Delta p = -\rho gh$, where Δp is the change in fluid pressure corresponding to an increase in height h, ρ is the density of the fluid and g is the gravitational acceleration.   If ρ is expressed in grams per cubic centimeter, g in centimeters per second per second and h in centimeters, will Δp be expressed in pascals (newtons per square meter)?

Let us first establish the fact that the equation is dimensionally consistent.   If we substitute the dimensions for each term, we obtain

$$[M/LT^2] = [M/L^3] \cdot [L/T^2] \cdot [L] \qquad \text{or} \qquad [M/LT^2] = [M/LT^2]$$

(see Table 2-2). Hence the equation is dimensionally consistent.

Now let us substitute the actual units into the equation.

$$[N/m^2] = [g/cm^3] \cdot [cm/s^2] \cdot [cm] \qquad \text{or} \qquad [N/m^2] = [g/cm \cdot s^2]$$

Now a newton is defined as 1 kg $\cdot$ m/s^2 (Table 2-4).   Substituting this definition into the left side of the equation and simplifying, we obtain

$$[kg/m \cdot s^2] = [g/cm \cdot s^2]$$

These units are not consistent.   Hence Δp will not be expressed in pascals.

If consistent SI units had been chosen for this example, the density would be expressed in kg/m^3, the gravitational acceleration in m/s^2 and the height in m.   The corresponding unit for Δp would then be pascals.

The requirement that an equation have consistent units allows us to determine the proper units for some particular term in an equation.

Example 2.29.   The work done by an ideal gas in expanding from volume V_1 to volume V_2 is given by the equation

$$W = nRT \ln (V_2/V_1)$$

where n represents the number of moles of gas, R is the *universal gas constant* and T is the absolute temperature of the gas (assumed constant).   If the temperature is expressed in kelvins and the work is to be expressed in joules, in what units should the universal gas constant be expressed?

If we solve for R and write the appropriate units for each variable (see Tables 2-3 and 2-4), we obtain

$$R = \frac{W}{nT \ln (V_2/V_1)} \qquad \frac{[J]}{[mol] \cdot [K]}$$

(Remember that $\ln(V_2/V_1)$ is a dimensionless quantity.)   Hence the desired units for the universal gas constant are J/mol $\cdot$ K.   The proper value of R, expressed in these units, is $R = 8.314$ J/mol $\cdot$ K.

If an equation contains a sum (or difference) involving two or more terms, then each term must be expressed in the same units.

Example 2.30.   The pressure drop for a fluid flowing through a pipe can be expressed as

$$\frac{\Delta p}{L} = av + bv^2$$

where Δp = the change in pressure, expressed in newtons per square meter
          L = the length of pipe over which the pressure drop occurs, expressed in meters
          v = the fluid velocity, in meters per second
In what units should the constants a and b be expressed?

The left side of the equation has the units   [N/m^2] $\div$ [m] = [N/m^3].   Therefore each term on the right side of the equation must also have the units [N/m^3].   Since velocity has the units [m/s], we can write

$$[av] = [a] \cdot [m/s] = [N/m^3]$$

Therefore the units of a must be

$$[a] = [N/m^3] \div [m/s] = [N \cdot s/m^4]$$

Similarly,

$$[bv^2] = [b] \cdot [m^2/s^2] = [N/m^3] \qquad \text{or} \qquad [b] = [N \cdot s^2/m^5]$$

Solved Problems

2.1. A units system is based upon the following fundamental dimensions: $[L]$, $[M]$, $[T]$, $[Q]$, $[\theta]$, $[I]$. What will be the proper symbolic representation for (*a*) momentum, (*b*) pressure, (*c*) energy, (*d*) power, (*e*) density, (*f*) viscosity, (*g*) thermal conductivity, (*h*) electric potential, (*i*) electric resistance, (*j*) electric capacitance?

> The proper symbols can be found in Table 2-2.
> (*a*) $[ML/T]$ (*c*) $[ML^2/T^2]$ (*e*) $[M/L^3]$ (*g*) $[ML/T^3\theta]$ (*i*) $[ML^2/Q^2T]$
> (*b*) $[M/LT^2]$ (*d*) $[ML^2/T^3]$ (*f*) $[M/LT]$ (*h*) $[ML^2/QT^2]$ (*j*) $[Q^2T^2/ML^2]$

2.2. In an *absolute* units system, force is a derived dimension that is defined in terms of mass and acceleration. What is the proper unit of force in the absolute metric cgs (*centi-meter-gram-second*) system?

> From Newton's second law, $F = ma$,
>
> $$[F] = [M] \cdot [L/T^2] = [g] \cdot [cm/s^2]$$

Hence the unit of force is the g $\cdot$ cm/s², which is called the *dyne*.

2.3. In a *gravitational* units system, mass is a derived dimension that is defined in terms of force and acceleration. What is the proper unit of mass in the gravitational English fps system?

> From Newton's second law, $F = ma$ or $m = F/a$,
>
> $$[M] = \frac{[F]}{[L/T^2]} = \frac{[lb_f]}{[ft/sec^2]}$$

Hence the unit of mass is the $lb_f \cdot sec^2/ft$, which is called the *slug*.

2.4. A force of 4 N is required to accelerate a body at 20 m/s². What is the mass of the body, expressed in grams?

$$m = \frac{F}{a} = \frac{4 \text{ N}}{20 \text{ m/s}^2} = 0.2 \text{ kg} = 200 \text{ g}$$

2.5. What will be the weight of a 75 kg person on a space satellite, where the gravitational acceleration (simulated by rotation) is 80 percent of the value on earth? Express the answer in SI units.

> The gravitational acceleration on earth (at sea level and a latitude of 45°) is 9.807 m/s². Hence the gravitational acceleration on the space satellite will be (0.8)(9.807) = 7.8456 m/s². The person's weight will therefore be
>
> $$(75 \text{ kg})(7.8456 \text{ m/s}^2) = 588.42 \text{ N}$$

2.6. A metal cube 2 m on a side, whose mass is 5000 kg, rests on a flat surface. Determine the pressure exerted on the surface, in SI units.

> The weight of the cube will be
>
> $$(5000 \text{ kg})(9.807 \text{ m/s}^2) = 4.9035 \times 10^4 \text{ N}$$

The area of one face of the cube is 4 m². Hence the pressure exerted by the cube is

$$\frac{4.0935 \times 10^4 \text{ N}}{4 \text{ m}^2} = 1.226 \times 10^4 \text{ Pa}$$

2.7. How much work is done if a force of 25 N acts through a distance of 6 dm?

Work is defined as: force $\times$ distance. Since 6 dm = 0.6 m, we can write

$$(25 \text{ N})(0.6 \text{ m}) = 15 \text{ J}$$

2.8. What is the power output corresponding to an energy consumption of 10^3 kcal/h?

Since 10^3 kcal = 4184×10^3 J and 1 hr = 3600 s, the power output will be

$$4184 \times 10^3 \text{ J}/ 3600 \text{ s} = 1.162 \times 10^3 \text{ W} = 1.162 \text{ kW}$$

2.9. What is the power output corresponding to a current of 115 A and a potential of 2000 V?

Power can be defined as: current $\times$ potential. Hence

$$(115 \text{ A})(2000 \text{ V}) = 230 \times 10^3 \text{ W} = 230 \text{ kW}$$

2.10. A current of 30 mA flows across a 10 Ω resistance. What will be the corresponding voltage drop?

Voltage can be expressed as the product of current and resistance. Hence

$$(30 \times 10^{-3} \text{ A})(10 \text{ }\Omega) = 0.30 \text{ V}$$

2.11. What will be the weight, in poundals and in lb_f, of a 160 lb_m man?

Under standard earth gravity a mass of 1 lb_m weighs 32.174 poundals or 1 lb_f. Hence, 160 lb_m weighs

$$160(32.174) = 5147.8 \text{ poundals} \qquad \text{or} \qquad 160(1) = 160 \text{ lb}_f$$

Alternatively, using Newton's second law,

$$F = \frac{ma}{g_c} = \frac{(160 \text{ lb}_m)(32.174 \text{ ft/sec}^2)}{32.174 \text{ lb}_m \cdot \text{ft/lb}_f \cdot \text{sec}^2} = 160 \text{ lb}_f$$

2.12. What is the mass of an object that would weigh 1 lb_f on Jupiter, where the gravitational acceleration is 2.65 times that on earth?

From Newton's second law,

$$m = \frac{g_c F}{a} = \frac{(32.174 \text{ lb}_m \cdot \text{ft/lb}_f \cdot \text{sec}^2)(1 \text{ lb}_f)}{(2.65)(32.174 \text{ ft/sec}^2)} = 0.377 \text{ lb}_m$$

2.13. How many tons of coal are contained in 1 acre of a coal seam that is 6 ft thick, if the density of the coal is 90 lb_m/ft^3?

From Table 2-8, 1 acre = 43 560 ft^2. Hence the mass of coal is

$$(43 \text{ }560 \text{ ft}^2)(6 \text{ ft})(90 \text{ lb}_m/\text{ft}^3) = 23 \text{ }522 \text{ }400 \text{ lb}_m$$

Converting to tons,

$$(23 \text{ }522 \text{ }400 \text{ lb}_m)\left(\frac{1 \text{ ton}}{2000 \text{ lb}_m}\right) = 11 \text{ }761.2 \text{ tons}$$

2.14. A force of 200 pounds acts against a steel plate whose dimensions are 5 yd by 10 in. Determine the pressure, in atmospheres.

The area of the plate is

$$[(5 \text{ yd})(3 \text{ ft/yd})]\left[(10 \text{ in})\left(\frac{1 \text{ ft}}{12 \text{ in}}\right)\right] = 12.5 \text{ ft}^2$$

Hence the pressure will be

$$(200 \text{ lb}_f) \div (12.5 \text{ ft}^2) = 16 \text{ lb}_f/\text{ft}^2$$

Converting to atmospheres,

$$(16 \text{ lb}_f/\text{ft}^2)\left(\frac{1 \text{ atm}}{2116.224 \text{ lb}_f/\text{ft}^2}\right) = 7.560\,64 \times 10^{-3} \text{ atm}$$

2.15. How many Btu per second are equivalent to 1 horsepower?

From Table 2-8, 1 hp = 550 ft · lb$_f$/sec, and 1 Btu = 777.65 ft · lb$_f$. The latter units equivalence can also be written as 1 Btu/sec = 777.65 ft · lb$_f$/sec. Hence,

$$(1 \text{ hp})\left(\frac{550 \text{ ft} \cdot \text{lb}_f/\text{sec}}{1 \text{ hp}}\right)\left(\frac{1 \text{ Btu/sec}}{777.65 \text{ ft} \cdot \text{lb}_f/\text{sec}}\right) = 0.707\,259 \text{ Btu/sec}$$

Thus we see that 1 hp is equivalent to approximately 0.7 Btu/sec.

Notice the way the units cancel each other in this calculation, leaving only the desired units, Btu/sec.

2.16. Convert 25 kg/cm² to lb$_m$/in².

From Appendix A,

$$(25 \text{ kg/cm}^2)\left(\frac{2.54 \text{ cm}}{1 \text{ in}}\right)^2\left(\frac{2.204\,62 \text{ lb}_m}{1 \text{ kg}}\right) = (25 \text{ kg/cm}^2)\left(\frac{6.4516 \text{ cm}^2}{1 \text{ in}^2}\right)\left(\frac{2.204\,62 \text{ lb}_m}{1 \text{ kg}}\right)$$

$$= 355.583 \text{ lb}_m/\text{in}^2$$

2.17. Convert 200 hp to J/h.

Note that 1 J/s = 1 W (by definition). Hence, from Appendix A,

$$(200 \text{ hp})\left(\frac{745.7 \text{ J/s}}{1 \text{ hp}}\right)\left(\frac{3600 \text{ s}}{1 \text{ h}}\right) = 5.369\,04 \times 10^8 \text{ J/h}$$

2.18. Convert a temperature reading of 650 °R to an equivalent temperature in °C.

From Appendix A,

$$\theta_F = 1.8\theta_C + 32 = \theta_R - 459.67$$

Hence

$$\theta_C = \frac{\theta_R - 491.67}{1.8} = \frac{650 - 491.67}{1.8} = 87.96 \text{ °C}$$

2.19. Convert a temperature difference of 25 °C to an equivalent temperature difference in degrees Fahrenheit.

From Appendix A 1.8 °F = 1 °C. Hence,

$$(25 \text{ °C})\left(\frac{1.8 \text{ °F}}{1 \text{ °C}}\right) = 45 \text{ °F}$$

Comparing this problem with Problem 2.18 we see that the conversion of a *temperature difference* is calculated differently than the conversion of a *temperature reading*. A units equivalence is used in the former case, and a temperature conversion formula in the latter.

2.20. Convert 20 centipoise (viscosity) to $lb_m/ft \cdot hr$.

$$(20 \text{ centipoise})\left(\frac{2.4191 \text{ } lb_m/ft \cdot hr}{1 \text{ centipoise}}\right) = 48.382 \text{ } lb_m/ft \cdot hr$$

2.21. The shear stress in a beam can be determined by the formula

$$\tau = \frac{VQ}{bI}$$

where τ is the shear stress (force per unit area), V is the external force, Q is a geometric factor having the dimension of volume, b is the thickness of the beam, and I is the *moment of inertia* (of the beam's cross-sectional area). What are the proper dimensions for the moment of inertia?

Solving for I, we obtain $I = VQ/b\tau$, and so

$$[I] = \frac{[V] \cdot [Q]}{[b] \cdot [\tau]} = \frac{[F] \cdot [L^3]}{[L] \cdot [F/L^2]} = [L^4]$$

2.22. The motion of a vibrating mass can be determined by the equation

$$x = e^{-\alpha t}(C_1 \sin \beta t + C_2 \cos \beta t)$$

where x is the distance from the equilibrium position at any time t. Determine the units of α, β, C_1 and C_2 if x is expressed in meters and t in seconds.

The arguments of the exponential and trigonometric functions must be dimensionless. Therefore α and β must each be expressed in terms of s^{-1}. Also, the exponential and trigonometric functions are themselves dimensionless. Hence C_1 and C_2 must each be expressed in meters.

Supplementary Problems

Answers are provided at the end of the book.

2.23. Express the fundamental dimensions of SI in symbolic notation.

2.24. Express the fundamental dimensions of the American engineering system in symbolic notation.

2.25. A units system is based upon the fundamental dimensions $[L]$, $[T]$, $[F]$, $[A]$, $[\theta]$, $[I]$. What will be the proper symbolic representation for each of the following derived dimensions: (a) momentum, (b) stress, (c) energy, (d) power, (e) density, (f) viscosity, (g) thermal conductivity, (h) electric potential, (i) electric resistance, (j) electric inductance?

2.26. In an *absolute* units system, force is a derived dimension that is defined in terms of mass and acceleration. What is the proper unit of force in the absolute English fps system?

2.27. In a *gravitational* units system, mass is a derived dimension that is defined in terms of force and acceleration. What is the proper unit of mass in the gravitational metric mks system, where the unit of force is the kg_f?

2.28. Define the following SI units: (*a*) newton, (*b*) joule, (*c*) watt, (*d*) hertz, (*e*) pascal, (*f*) coulomb, (*g*) volt, (*h*) ohm, (*i*) farad, (*j*) henry, (*k*) weber, (*l*) tesla, (*m*) lumen, (*n*) lux.

2.29. Define the following SI prefixes: (*a*) centi, (*b*) deci, (*c*) kilo, (*d*) mega, (*e*) micro, (*f*) milli, (*g*) nano, (*h*) pico.

2.30. Determine the force required for a mass of 30 g to be accelerated 180 cm/s².

2.31. A 12 kg mass weighs 54 N on the surface of a distant planet. What is the gravitational acceleration on that planet?

2.32. An object resting on a flat surface exerts a pressure of 10 kPa on the surface. If the surface has an area of 4 cm², what is the mass of the object?

2.33. How much work is required to raise a 50 kg mass 80 cm?

2.34. How many kcal/s are required to power a 100 W light bulb?

2.35. What is the power output corresponding to an energy consumption of 10^7 N · m/h?

2.36. If 3.5 kg of water is supplied with 50 kJ in the form of heat, how large will be the resulting temperature increase?

2.37. An electronic device has a power rating of 0.5 W. If the device consumes 20 mA, what is the corresponding voltage drop?

2.38. If the device in Problem 2.37 is a resistor, what is its resistance?

2.39. What charge is required to produce a current of 20 A for one minute?

2.40. What is the capacitance corresponding to a charge of 8 C acting across a potential difference of 20 V?

2.41. How many μs are there in 1 min?

2.42. How many mg are there in 1 kg?

2.43. One ms is equivalent to how many ps?

2.44. A modern computer can add two numbers in about 100 ns. How many additions can be carried out in the time required to blink an eye (about 0.1 s)?

2.45. A man weighs 175 pounds (on earth, at sea level and a latitude of 45°). (*a*) What is his weight in poundals? (*b*) What is his mass? (*c*) What will he weigh, in lb_f and in poundals, on a distant planet where the gravitational acceleration is double the value on earth? (*d*) What will be his mass on the distant planet?

2.46. What acceleration will result if a force of 150 lb is applied to a 40 lb mass?

2.47.   How much work is required to accelerate a mass of 0.2 ton at 4 ft/sec² through a distance of 6 ft?

2.48.   How much water is contained in a 50 gal container?   (The density of water is 62.4 lb$_m$/ft³.)

2.49.   What force is required to produce a pressure of 75 atm, if the force acts against a steel plate whose dimensions are 2 ft × 4 in?

2.50.   How many ft · lb$_f$/sec are equivalent to 150 hp?

2.51.   How many horsepower are equivalent to 375 Btu/sec?

2.52.   If 128 Btu is supplied to 8.3 lb$_m$ of water originally at 58.7 °F, what will be the final temperature of the water?

2.53.   Carry out the following units conversions: (*a*)   156 kg to lb$_m$,   (*b*)   47.2 lb$_m$ to g,   (*c*)   0.0713 ton to kg,   (*d*)   625 ft to m,   (*e*)   39.2 cm to in,   (*f*)   0.001 in to Å,   (*g*)   200 mi to km,   (*h*)   72 μm to ft,   (*i*)   0.033 min to μs,   (*j*)   12 ns to ps,   (*k*)   58.1 ms to h,   (*l*)   800 N to lb$_f$,   (*m*)   53.9 poundals to dynes,   (*n*)   0.887 lb$_f$ to N,   (*o*)   208 Btu to kcal,   (*p*)   208 Btu to J,   (*q*)   125 000 J to hp · hr,   (*r*)   125 000 J to kW · h,   (*s*)   5 × 10⁶ kcal to kW · hr,   (*t*)   700 hp to kW,   (*u*)   425 W to hp,   (*v*)   2000 kcal/s to MW,   (*w*)   17 L to gal,   (*x*)   0.667 ft³ to L,   (*y*)   5 acre · ft to m³,   (*z*)   8 atm to Pa.

2.54.   Is a degree Rankine larger or smaller than a kelvin?

2.55.   Several temperature readings (temperature measurements) are given below.   Carry out the indicated conversions.

        (*a*)   165 °F to °R, °C and K   (*c*)   1200 °R to K, °F and °C
        (*b*)   25 °C to K, °F and °R   (*d*)   220 K to °C, °R and °F

2.56.   Several temperature differences are given below.   Carry out the indicated conversions.

        (*a*)   120 °C to K and °R   (*c*)   1000 K to °F and °C
        (*b*)   45 °F to °C and K   (*d*)   270 °R to °F and K

2.57.   Carry out the following units conversions.

        (*a*)   105 lb$_m$/ft³ to g/cm³   (*g*)   750 Btu/hr · ft² to kcal/s · cm²
        (*b*)   0.915 g/cm³ to lb$_m$/ft³   (*h*)   30 lb$_f$/in to N/cm
        (*c*)   4.4 × 10⁵ Btu/hr to kcal/s   (*i*)   4.08 lb$_m$/min · ft² to kg/s · m²
        (*d*)   4.4 × 10⁵ Btu/hr to J/min   (*j*)   12 500 gal/hr to L/min
        (*e*)   810 kcal/min to ft · lb$_f$/sec   (*k*)   15 Btu/hr · ft² · °F to W/m² · K
        (*f*)   3500 kcal/m² to Btu/ft²   (*l*)   6.67 ft²/hr to stokeses (kinematic viscosity)

2.58.   A copper wire has a diameter of 2 mm.   If the density of copper is 8300 kg/m³, what will be the weight of 100 m of the wire?

2.59.   A steel I-beam weighs 225 lb$_f$ per ft of length.   What is the cross-sectional area of the I-beam, in square inches, if the density of steel is 440 lb$_m$/ft³?

2.60.   A river discharges 160 000 gal/sec of water at its mouth.   If the cross section of the river bed has the shape of a semicircle with a diameter of 1/8 mi, what will be the velocity of the water flow?   Express your answer in ft/sec and km/h.

2.61. Determine whether or not each of the following equations is dimensionally consistent.

(a)

$$h = \sqrt{\frac{\mu x}{\rho v}}$$

where h = height of fluid layer
         μ = viscosity of fluid
         x = distance
         ρ = fluid density
         v = fluid velocity

(b)

$$v = \sqrt{2cgT\,[1-(p_2/p_1)^{(k-1)/k}]}$$

where v = gas velocity
         c = heat capacity
         g = gravitational acceleration
         T = absolute temperature

(c)

$$I = V\left[R^2 + \left(2\pi fL - \frac{1}{2\pi fC}\right)^2\right]^{-1/2}$$

where I = electric current
         V = voltage (potential)
         R = resistance
         f = frequency
         L = inductance
         C = capacitance

2.62. In fluid mechanics, the frictional force of a fluid acting against a smooth surface is given by the formula

$$F = \frac{1}{2}\,c\rho a v^2$$

where F = frictional force
         c = coefficient of friction
         ρ = fluid density
         a = surface area
         v = fluid velocity

Determine the dimensions for c, the coefficient of friction.

2.63. The volume of a substance at temperature T can be calculated from the formula

$$V = V_0(1 + \beta T)$$

where V = volume at temperature T
         V_0 = volume at temperature zero
         β = coefficient of volumetric expansion

(a) Determine the dimensions for β.   (b) Determine the proper units for β if V is expressed in liters and T in kelvins.

2.64. The temperature, pressure and volume of a gas are related by the formula

$$p = \frac{RT}{V-a} - \frac{b}{V^2}$$

where p = gas pressure
         R = universal gas constant
         T = absolute gas temperature
         V = gas volume, per mole
       a, b = corrective constants

(a) Determine the dimensions for R, a, and b.   (b) Determine the proper SI units for R, a, and b.   (c) Determine the proper English units for R, a, and b (assume that p is expressed in atmospheres).

2.65. The load that will cause a column to buckle (collapse) is given by the formula

$$F = \frac{\pi^2 \, EI}{L^2}$$

where F = force (load) that results in buckling
  E = Young's modulus
  I = cross-sectional moment of inertia
  L = column length

(a)   Determine the dimensions for E if I has the dimensions $[L^4]$.   (b)   If I is expressed in m⁴, what are the proper units for E?   (c)   If I is expressed in in⁴, determine the proper units for L and for E.

2.66. The rate at which heat is transferred across a surface is determined by the formula

$$Q = UA \, \Delta T$$

where Q = heat transfer rate
  U = heat transfer coefficient
  A = surface area
  ΔT = temperature difference across the surface

(a)   If the heat transfer rate is expressed as energy per unit time, what are the required dimensions for U?   (b)   What are the proper SI units for U?   (c)   What are the proper units for U if Q is expressed in Btu/hr and A is expressed in ft²?

Chapter 3

Graphs and Nomographs

A *graph* allows an equation or a data set to be pictured on a two-dimensional grid, whereas a *nomograph* is used to represent an equation by means of several carefully positioned scales. These devices reveal relationships that might otherwise be obscured by complicated algebraic equations or long tables of numbers.

3.1 CARTESIAN COORDINATES

The most common type of graph is based upon the use of *cartesian coordinates* (also called *arithmetic coordinates* or *rectangular coordinates*). When data are represented in this manner, the dependent variable is plotted along the vertical axis (the *ordinate*) and the independent variable is plotted along the horizontal axis (the *abscissa*). Each axis should be clearly labeled, indicating the scale, the variable being plotted and its units.

Example 3.1. The position of a falling object can be determined as a function of time by the equation

$$y = 200 - 4.9t^2$$

where y = height of the object above the ground, in meters
t = time since the object was released, in seconds

Prepare a plot of y versus t.

It is easiest to begin by tabulating corresponding values of y and t, as shown below. These values can then be plotted.

t, s	0	1	2	3	4	5	6	6.39
y, m	200	195.1	180.4	155.9	121.6	77.5	23.6	0.0

The last value was obtained by solving the given equation for the value of t corresponding to $y = 0$.

The above data have been plotted in Fig. 3-1 and a smooth curve drawn through the individual data points.

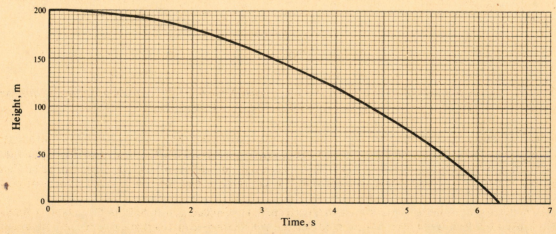

Fig. 3-1

44

If the data represent measured rather than calculated quantities, then it may be desirable to show the actual data points on the graph. This can easily be accomplished by enclosing each data point within a circle, triangle, square, etc. A smooth curve can then be drawn through the aggregate of data. This curve need not, however, pass through every data point.

Example 3.2. A student has carried out an experiment to study the chemical reaction $A + B \rightarrow C$. His measurements for $[C]$, the concentration of C, in moles per liter, at various times are as follows.

t, s	0	0.5	1.0	1.5	2.0	2.5	3.0	3.5
$[C]$, mol/L	0	0.05	0.24	0.40	0.60	0.95	1.24	1.41
t	4.0	4.5	5.0	5.5	6.0	6.5	7.0	
$[C]$	1.64	1.80	1.89	1.96	2.03	2.04	2.05	

From these data, construct a smoothed curve showing the variation of $[C]$ with time.

The data are plotted in Fig. 3-2. Notice that each data point is surrounded by a circle. A continuous curve, showing the approximate variation of concentration with time, has been drawn through the data points. In order to smooth out the data, however, the curve does not necessarily pass through each data point. Rather, the curve attempts to represent the general trend indicated by the data.

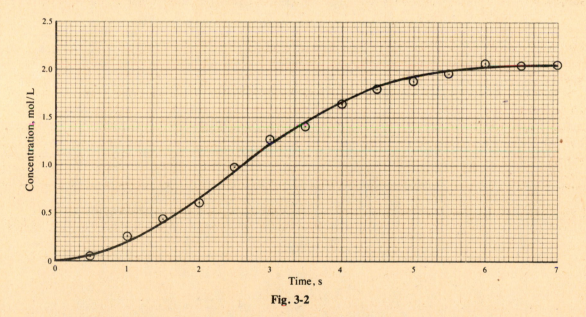

Fig. 3-2

Sometimes several sets of data are plotted on the same graph. In such situations it is important that each set of data be clearly identified and distinct from all the others.

Example 3.3. Several tests have been conducted on a structural component whose cumulative probability of failure increases markedly with temperature once a certain critical temperature has been exceeded. The results of three tests are presented in Table 3-1. Each test was carried out with a different material, having a different critical temperature. For purposes of comparison, it is desirable to plot the results of all three tests on the same graph.

The data are plotted in Fig. 3-3. Notice that each curve is identified as test A, B, or C, and that a different symbol is used for each set of data. Also, note that the temperature scale (plotted along the abscissa) begins at 1200 K rather than at 0 K.

Table 3-1

Test A		Test B		Test C	
Temperature, K	Cumulative Probability of Failure, %	Temperature, K	Cumulative Probability of Failure, %	Temperature, K	Cumulative Probability of Failure, %
1220	0	1250	0	1280	0
1225	2	1255	1.5	1285	2
1230	4	1260	3	1290	4
1235	12	1265	7	1295	8
1240	20	1270	13	1300	12
1245	35	1275	22	1305	17
1250	55	1280	30	1310	25
1255	75	1285	44	1315	32
1260	85	1290	55	1320	40
1265	94	1295	65	1325	48
1270	98	1300	75	1330	58
1275	99.5	1305	81	1335	67
1280	100	1310	88	1340	74
		1315	94	1345	80
		1320	97	1350	86
		1325	99	1355	91
		1330	100	1360	95
				1365	98
				1370	99
				1375	100

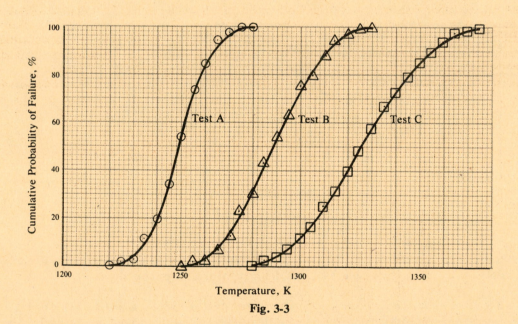

Fig. 3-3

A graph need not be restricted to positive sets of data. Both positive and negative values can be plotted along each axis, thus causing the graph to be divided into four *quadrants*. The upper-right quadrant, which contains only positive data points, is referred to as the *first quadrant* (quadrant I). The other quadrants are numbered consecutively by proceeding counterclockwise from quadrant I.

Example 3.4.   A mass resting on a horizontal surface is held between two horizontal springs.   If the mass is displaced in the positive x-direction (i.e. toward the right), then the springs exert a force in the opposite direction (to the left), which tends to restore the mass to its original (rest) position.   The same thing happens if the mass is displaced in the negative x-direction, producing a restoring force in the positive x-direction.

Figure 3-4 shows a plot of force versus displacement.   Notice that the curve appears in the second (upper-left) and fourth (lower-right) quadrants.   The reason, of course, is that a positive displacement produces a negative force, and a negative displacement produces a positive force.

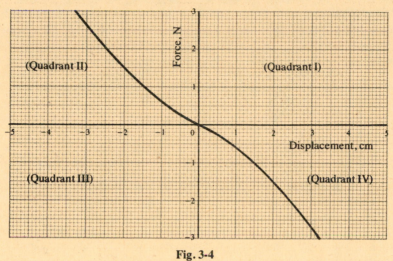

Fig. 3-4

It is very important that the divisions along each axis be clear and easy to read.   Complicated or obscure scale factors should be avoided.

Example 3.5.   Figure 3-5 shows two ways of labeling the ordinate in the graph of a certain set of stress-strain data.   Suppose that we wanted to determine the stress value corresponding to a strain of 0.001 75.   The ordinate value is 350.   Thus, according to the left-hand labeling, the stress $\times 10^{-7}$ is 350 N/m², so that the actual stress is 350×10^7 N/m².   It would be very easy, however, to misinterpret the scale factor along the ordinate, and conclude (incorrectly) that the stress corresponding to a strain of 0.001 75 is 350×10^{-7} N/m².

The right-hand labeling is a much better one.   It clearly shows that the stress value corresponding to a strain of 0.001 75 is $350 \times (10^7$ N/m²), or 350×10^7 N/m².

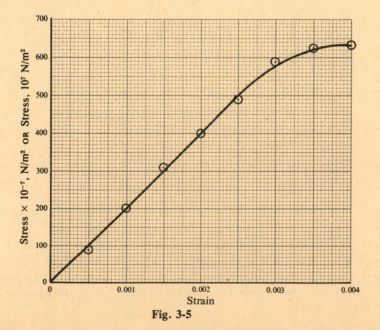

Fig. 3-5

3.2   BAR GRAPHS

When data are actually being measured, the values of the dependent variable may correspond to *discrete intervals* of the independent variable rather than individual points.   In such cases it may be desirable to plot the data in the form of vertical rectangles that are adjacent to one another.   This type of representation is known as a *bar graph*.

Example 3.6.   In Table 3-2 is tabulated a manufacturer's output of single-lens reflex cameras, on a monthly basis, for the past calendar year.

Table 3-2

Month	Number of Cameras	Month	Number of Cameras
January	8774	July	8831
February	9308	August	9147
March	9760	September	9382
April	9644	October	9549
May	9215	November	9130
June	9058	December	8983

The data are shown plotted in the form of a bar graph in Fig. 3-6.   This type of graph is appropriate because of the month-to-month fluctuations in the data.   These fluctuations would be less apparent if the data were represented by a smooth curve.   (Smoothing out the fluctuations would be undesirable in this situation.)

Notice that the ordinate in Fig. 3-6 begins with a value of 8000 rather than zero.   This causes the scale to be magnified, which again emphasizes the fluctuations in the data.

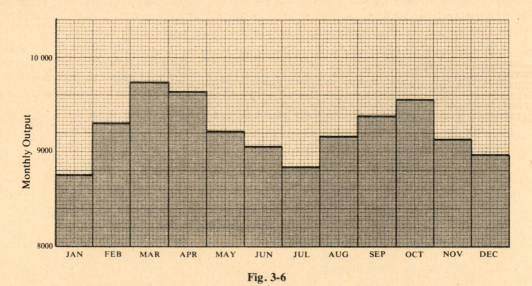

Fig. 3-6

A type of bar graph that is particularly important in statistical applications is the *histogram*.   This is a plot of the *relative frequency* of an event within each of several adjacent intervals.

Example 3.7.   In manufacturing an automobile engine, the cylinders must each be machined to a 90 mm diameter.   The machining must be carried out as accurately as possible, although there will always be some small variation in the diameter of each cylinder.

In order to estimate the degree of quality control on a given day, 200 engine blocks are selected at random after they have been machined; and the diameter of one of the cylinders in each engine block is accurately measured.   The collected results for a certain day are given in the first two columns of Table 3-3.

Table 3-3

Interval, mm	Number of Samples	Relative Frequency, %	Cumulative Relative Frequency, %
89.1–89.3	2	1.0	1.0
89.3–89.5	5	2.5	3.5
89.5–89.7	11	5.5	9.0
89.7–89.9	29	14.5	23.5
89.9–90.1	107	53.5	77.0
90.1–90.3	27	13.5	90.5
90.3–90.5	9	4.5	95.0
90.5–90.7	7	3.5	98.5
90.7–90.9	3	1.5	100.0
TOTALS	200	100.0%	

Thus 2 of the measured diameters were between 89.1 and 89.3 mm, 5 measurements were between 89.3 and 89.5 mm, and so on.   Notice that all of the measurements fell within ±1 mm of the desired 90 mm value.

The relative frequency (the third column of Table 3-3) can easily be determined by dividing the number of samples for each interval by the total sample size, and then multiplying by 100 in order to convert to a percentage.   Thus the relative frequency for the interval 89.1–89.3 mm is computed as

$$\text{relative frequency} = \frac{2}{200} \times 100 = 1\%$$

and so on.

A histogram of the relative frequencies is shown in Fig. 3-7.   Notice that the abscissa is divided at the midpoints of the intervals rather than at the interval boundaries.   This is not necessary, but it is fairly common, as the midpoint is usually assumed to be representative of the entire interval.

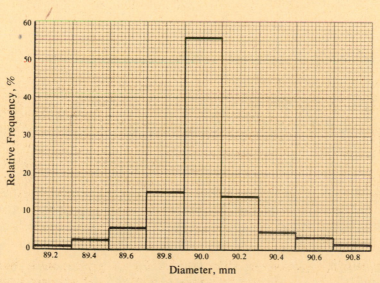

Fig. 3-7

Example 3.8.   Plot the *cumulative relative frequencies* for the data of Example 3.7.   (This is also called a *cumulative relative distribution* or simply a *cumulative distribution*.)

The cumulative relative frequencies are calculated in the fourth column of Table 3-3.   Each cumulative value consists of the sum of all relative frequencies from the first interval to the given interval.   In other words, the first cumulative relative frequency is simply 1.0; the second is   1.0 + 2.5 = 3.5;   and so on.

Figure 3-8 is a plot of the cumulative relative distribution. The abscissa is now divided at the interval boundaries rather than at the midpoints. This is quite common when plotting cumulative distribution data.

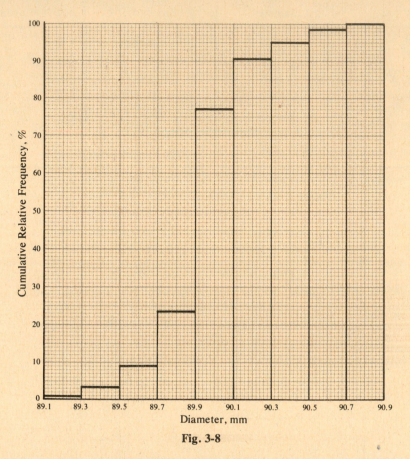

Fig. 3-8

We will say more about histograms and cumulative distribution graphs in Chapter 4.

3.3 SEMILOGARITHMIC COORDINATES

We have already noted that many scientific and technical calculations involve exponential equations of the form

$$y = ae^{bx}$$

(see Section 1.11). Moreover, we have seen that these equations generate the familiar exponential growth or exponential decay curves when plotted on ordinary arithmetic graph paper (see Fig. 1-1). An exponential equation can be represented as a straight line, however, by plotting ln y (or log y) against x. To see why this is so, take the natural logarithm of each side of the above equation, resulting in

$$\ln y = \ln ae^{bx} = \ln a + bx$$

Thus, if ln y is plotted against x, a straight line will be obtained having slope b and vertical intercept ln a.

For plotting purposes it is more convenient to use common logarithms (base 10) rather than natural logarithms. Since log $x \approx 0.434 \ln x$ (see Table 1-1),

$$\log y = \log a + (0.434b)x$$

Hence, if log y is plotted against x, we will obtain a straight line whose slope is $0.434b$ and whose vertical intercept is log a.

Example 3.9.   Draw a graph of the equation $y = 10e^{-0.5x}$ on cartesian coordinates.   Construct the graph two different ways: first plot y versus x, then log y versus x.

Several values of x, y, and log y, obtained by use of an electronic calculator, are given in Table 3-4.

Table 3-4

x	y	log y
0	10	1
1	6.07	0.783
2	3.68	0.566
3	2.23	0.349
4	1.35	0.131
5	0.821	−0.086
6	0.498	−0.303
7	0.302	−0.520
8	0.183	−0.737
9	0.111	−0.954

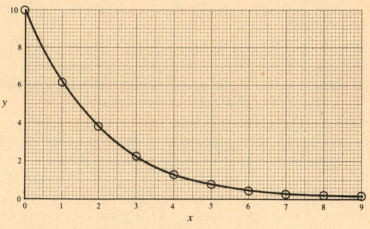

Fig. 3-9

The y-values are plotted in Fig. 3-9.   Notice the familiar exponential decay curve, which we would expect to obtain when plotting a negative exponential equation on arithmetic graph paper.

Figure 3-10 shows a plot of log y against x, again from Table 3-4.   Now we see that a straight line is obtained, whose vertical intercept is 1.0 and whose slope is −0.217.   From the given equation, $a = 10$ and therefore   log $a = 1$,   which corresponds to the value of the intercept as observed in Fig. 3-10.   Also, $b = -0.5$ in the given equation.   Hence the slope of the line should be

$$(0.434)(-0.05) = -0.217$$

as in fact it is in Fig. 3-10.

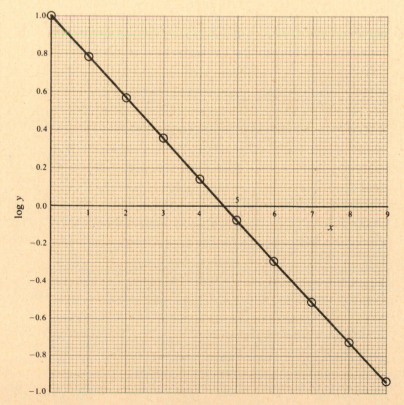

Fig. 3-10

A much easier way to plot log y against x is to use *semilogarithmic* graph paper, as shown in Fig. 3-11. In this type of graph the ordinate is divided into intervals proportional to log y rather than y, even though the ordinate is *labeled* in ordinary units of y. Thus it is not necessary to calculate the log of y when plotting an exponential equation; this is taken care of by the layout of the ordinate.

For example, the log of 2 is 0.301; therefore the ordinate value labeled "2" in Fig. 3-11 is located 30.1 percent of the way from the bottom. It is really log 2 that is located at this point, even though it is labeled "2." Notice that since log 1 = 0, the point labeled "1" is at the bottom of the ordinate. Also, log 10 = 1, so that the point labeled "10" is located 100 percent of the way from the bottom (i.e. at the top of the ordinate).

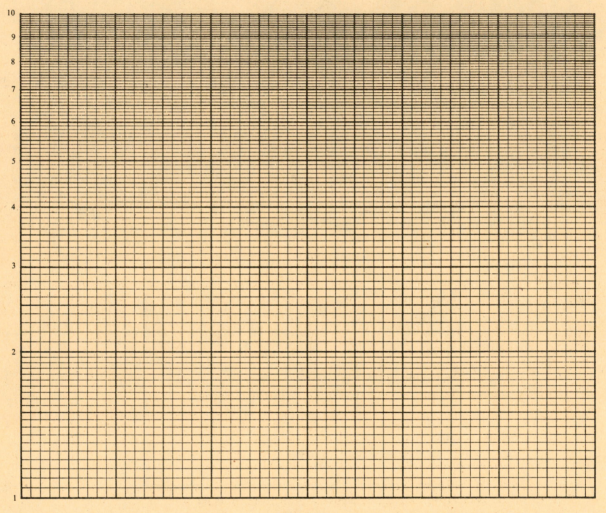

Fig. 3-11

Figure 3-11 is referred to as *one-cycle* semilog paper because the ordinate spans one order of magnitude (i.e. one factor of ten). Multiple-cycle semilog paper is also available, and is commonly used.

Example 3.10. Draw a graph of the equation $y = 10e^{-0.5x}$ on semilogarithmic coordinates.

We have already tabulated several values of y against x for this equation in Table 3-4. These values, plotted on semilog paper, result in the straight line shown in Fig. 3-12.

It should be understood that Fig. 3-12 is actually a plot of log y against x (compare with Fig. 3-10), even though we have plotted the tabulated values of y (not log y) versus x. Thus the conversion from y to log y is taken care of automatically by the semilogarithmic coordinates. Also, notice that we have utilized two-cycle semilog paper, since the y-values span two orders of magnitude.

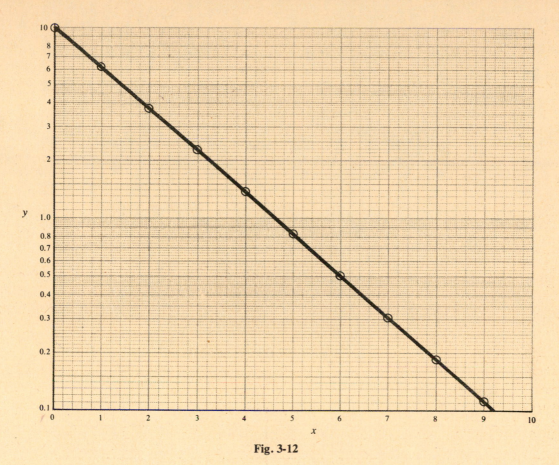

Fig. 3-12

Example 3.11. The rate at which a chemical reaction is carried out is determined by a *reaction rate constant*, k. This constant varies with temperature in accordance with the expression

$$k = Ae^{-E/RT}$$

where
  A = frequency factor, s^{-1}
  E = activation energy, J/mol
  R = universal gas constant, 8.32 J/mol · K
  T = absolute temperature, K

Determine the frequency factor and the activation energy for the following data.

T, K	338	328	318	308	298
k, 10^5 s^{-1}	475	150	50	13	3.5

The required information can be obtained by plotting k against $1/T$ on semilog paper. A straight line will be obtained whose slope is $-0.434E/R$ and whose vertical intercept is log A.

Figure 3-13 shows the data plotted on 3-cycle semilog paper. From this figure we see that the slope can be evaluated as

$$\frac{\Delta(\log k)}{\Delta(1/T)} = \frac{\log(1 \times 10^5) - \log(1000 \times 10^5)}{0.003\,46 - 0.002\,90} = \frac{5 - 8}{0.000\,56} = -5357$$

Hence

$$-0.434\frac{E}{R} = -5357 \qquad \text{or} \qquad E = \frac{(-5357)(8.32)}{-0.434} = 1.027 \times 10^5 \text{ J/mol}$$

The true vertical intercept cannot be read directly from Fig. 3-13 because the abscissa does not begin at zero. But because we have determined the value of E, we can calculate the value of A from

$$\log k = \log A - \left(0.434\frac{E}{R}\right)\frac{1}{T}$$

We know that $k = 1000 \times 10^5 = 10^8$ when $1/T = 0.0029$ (from Fig. 3-13).   Hence

$$\log 10^8 = \log A - (5357)(0.0029) \qquad \text{or} \qquad 8 = \log A - 15.5353$$

Solving for A, we obtain

$$\log A = 8 + 15.5353 = 23.5353 \qquad \text{or} \qquad A = 10^{23.5353} = 3.43 \times 10^{23} \text{ s}^{-1}$$

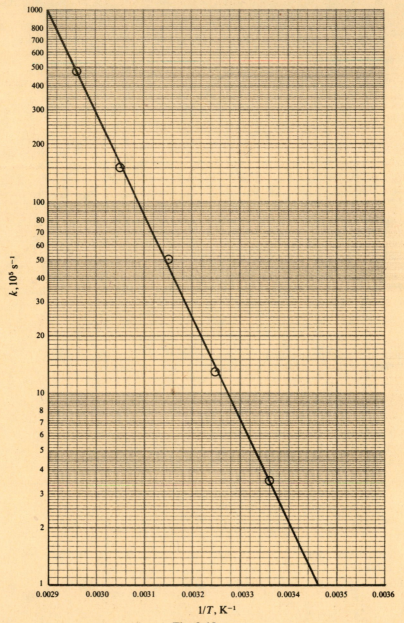

Fig. 3-13

   In order to check our results, let us see if we can back-calculate some other point on the line.   For example, we see that k should equal 7×10^5 when $1/T = 0.0033$ (i.e. $T = 303$ K).   If we calculate the value of k when $1/T = 0.0033$,   we obtain

$$k = (3.43 \times 10^{23})e^{-(1.027 \times 10^5/8.32)(0.0033)} = (3.43 \times 10^{23})e^{-40.734}$$
$$= (3.43 \times 10^{23})(2.0384 \times 10^{-18}) = 6.99 \times 10^5 \approx 7 \times 10^5$$

as expected.

3.4 LOGARITHMIC COORDINATES

Another equation that appears frequently in scientific and technical work is the *power function*

$$y = ax^b$$

This equation generates a curve when plotted on arithmetic graph paper (except for the special case where $b = 1$, which results in a straight line). A linear relationship can be obtained for any value of b, however, by plotting $\log y$ against $\log x$. In fact, taking the logarithm of each side of the above equation, we obtain

$$\log y = \log ax^b = \log a + b \log x$$

Therefore, if we plot $\log y$ against $\log x$, we will obtain a straight line whose slope is b and vertical intercept is $\log a$.

Example 3.12. Draw a graph of the equation $y = 0.2x^{1.8}$ on cartesian coordinates. Construct the graph two different ways: first plot y versus x, then $\log y$ versus $\log x$.

We begin by tabulating a number of corresponding values of x, y, $\log x$, and $\log y$, with the aid of an electronic calculator (see Table 3-5).

Table 3-5

x	y	$\log x$	$\log y$
0	0.0		
1	0.200	0.0	−0.699
2	0.696	0.301	−0.157
3	1.44	0.477	0.160
4	2.43	0.602	0.385
5	3.62	0.699	0.559
6	5.03	0.778	0.702
7	6.64	0.845	0.822
8	8.44	0.903	0.927
9	10.44	0.954	1.019

The tabulated values of y and x are plotted in Fig. 3-14. The curvature of this function is readily apparent.

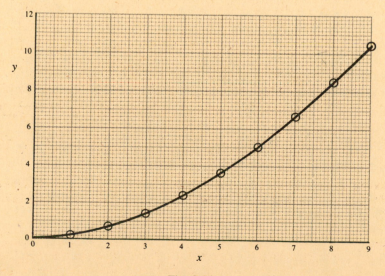

Fig. 3-14

Figure 3-15 shows a plot of log y versus log x, using the last two columns of Table 3-5. We now obtain a straight line having a slope of 1.80 and a vertical intercept of -0.699. If we examine the given equation, we see that $a = 0.2$. Thus

$$\log a = \log 0.2 = -0.699$$

as observed in Fig. 3-15. Also, $b = 1.80$ in the given equation, which corresponds to the value of the slope determined from Fig. 3-15.

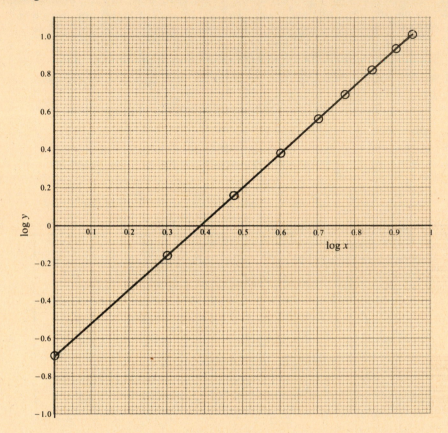

Fig. 3-15

A much easier way to plot log y against log x is to use *logarithmic coordinate* graph paper (commonly known as *log-log* paper), as shown in Fig. 3-16. This type of paper is similar to semilog paper, except that *both* the ordinate and the abscissa are divided into intervals that are proportional to log y and to log x. The axes are labeled in ordinary units of y or x, however, so that it is not necessary to actually calculate logarithms when plotting a power function on logarithmic coordinates.

Figure 3-16 is referred to as *1 × 1* (i.e. *1 cycle × 1 cycle*) log-log paper because the ordinate and the abscissa each span one order of magnitude (one factor of ten). Multiple-cycle log-log paper is also available.

Example 3.13. Draw a graph of the equation $y = 0.2x^{1.8}$ on log-log coordinates.

We have already obtained several values of y versus x for this equation in Table 3-5. These values can easily be plotted on 2 × 2 log-log paper, resulting in the straight line shown in Fig. 3-17.

Notice that the first tabulated point ($x = 0$, $y = 0$) cannot be plotted, since the ordinate and abscissa always begin at some nonzero value on a log-log plot. (In this particular case, the ordinate begins at $y = 0.1$ and the abscissa at $x = 1.0$.) Also, notice that the last tabulated point ($x = 9$, $y = 10.44$) extends slightly beyond the maximum ordinate value $y = 10$. Finally, note that all the tabulated x-values fall within one order of magnitude. Thus it probably would have been more convenient to plot the data on a 2 × 1, rather than a 2 × 2, log-log graph.

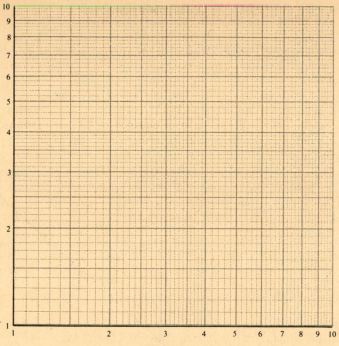

Fig. 3-16

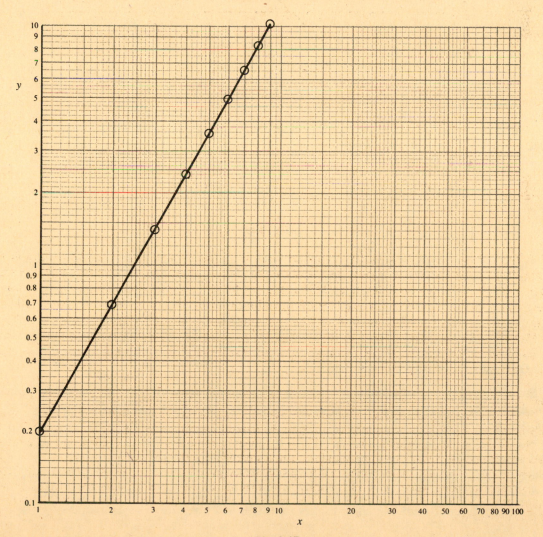

Fig. 3-17

The reader is again reminded that Fig. 3-17 is actually a plot of log y against log x (compare with Fig. 3-15), even though we have plotted the tabulated values of y versus x (not log y versus log x). Thus the conversion from y and x to log y and log x is taken care of automatically by the logarithmic coordinates.

Experimental data can very often be plotted as a straight line on log-log paper. This implies, of course, that such data can be represented by a power function. The constants in the power function can be determined graphically, by calculating the slope of the line that has been drawn through the data. When carrying out this calculation, however, it will be necessary to work with the logarithms of the values read from the ordinate and the abscissa, rather than the values themselves.

Example 3.14. Determine whether or not the following data can be represented by a power function. Also, evaluate the constants if the power function does appear valid.

x	2	4	7	10	20	40	60	80
y	45	26	17	12.5	7.3	4.2	3.0	2.4

Plotting the data on 2×2 log-log paper results in a well-defined straight-line relationship, as shown in Fig. 3-18. Thus the data can indeed be represented by a power function of the form

$$y = ax^b$$

The value of the constant b is equal to the slope of the line on the log-log plot. Thus, from Fig. 3-18,

$$b = \frac{\Delta(\log y)}{\Delta(\log x)} = \frac{\log 2 - \log 80}{\log 100 - \log 1} = \frac{0.301\,03 - 1.903\,09}{2 - 0} = -0.801\,03$$

We can make use of the known value of b to determine the value of a. To do so, let us write

$$\log y = \log a - 0.801\,03 \log x$$

We can now choose any point along the line, substitute the values for x and y into the above equation, and solve for log a. It is particularly convenient to choose the point corresponding to $x = 1$, since $\log 1 = 0$. From Fig. 3-18, we see that $y = 80$ when $x = 1$. Thus,

$$\log 80 = \log a - 0.801\,03 \log 1 = \log a$$

so that $a = 80$. The power function that corresponds to the straight line in Fig. 3-18 can therefore be written as

$$y = 80x^{-0.801\,03}$$

This result can easily be checked by back-calculating some other point on the line. For example, we see that $y = 2$ when $x = 100$. Substituting $x = 100$ into the above equation, we obtain

$$y = 80(100)^{-0.801\,03} = 80(0.025) = 2$$

as expected.

3.5 OTHER COORDINATE SYSTEMS

Several other types of graphs are used in technical applications, though much less commonly than arithmetic, semilog, and log-log graphs. In this section we will briefly discuss three such graphs.

Polar Coordinates

Some technical problems employ *polar coordinates*, where the dependent variable is represented by the radial distance from the origin. The independent variable is the angle between the radius and a fixed reference line; this angle ranges from 0 to 360° (or from 0 to 2π radians).

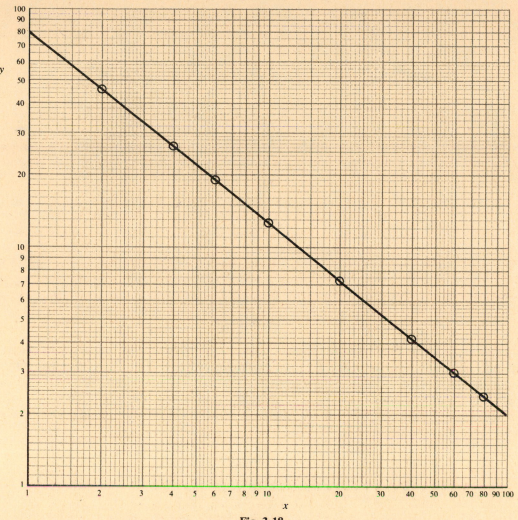

Fig. 3-18

Example 3.15. The temperature on the surface of a metal cylinder is given by the equation

$$T = 100 + 20 \cos (\theta/57.3) \qquad 0 \le \theta \le 360°$$

where T = temperature, °C (dependent variable)
θ = angle relative to some fixed position (independent variable)

Prepare a plot of T versus θ on arithmetic graph paper.

We begin by tabulating several values for θ and T, using an electronic calculator.

θ, deg	0	30	60	90	120	150	180
T, °C	120.0	117.3	110.0	100.0	90.0	82.7	80.0
θ	210	240	270	300	330	360	
T	82.7	90.0	100.0	110.0	117.3	120.0	

These values, when plotted, result in the smooth sinusoidal curve shown in Fig. 3-19.

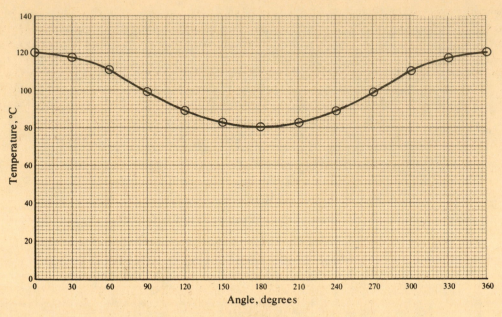

Fig. 3-19

When an equation is expressed in terms of polar coordinates it is sometimes convenient to plot the equation on *polar coordinate* graph paper, as shown in Fig. 3-20. The concentric circles form a scale for the dependent variable, with larger values of the dependent variable plotted farther from the origin than smaller values. The radii emanating outward from the origin are used to represent angular position. Notice that the radii are labeled in multiples of 10° from the lower vertical position (i.e. from the bottom of the graph).

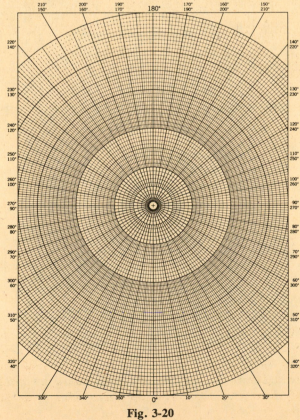

Fig. 3-20

Example 3.16.   Plot the equation given in Example 3.15 on polar coordinate graph paper.

Figure 3-21 shows a plot of the equation, based upon the tabulated values presented in Example 3.15.   Notice the circlelike, closed curve that has been drawn through the data points (although it is not a perfect circle), in contrast to the sinusoidal curve shown in Fig. 3-19.

Also, note that, because of symmetry, it has not been necessary to specify whether the angular measurements are clockwise or counterclockwise.   It is customary, however, for θ to increase in the counterclockwise direction.   This information must be specified (or understood) for those plots that are not symmetric about the vertical axis.

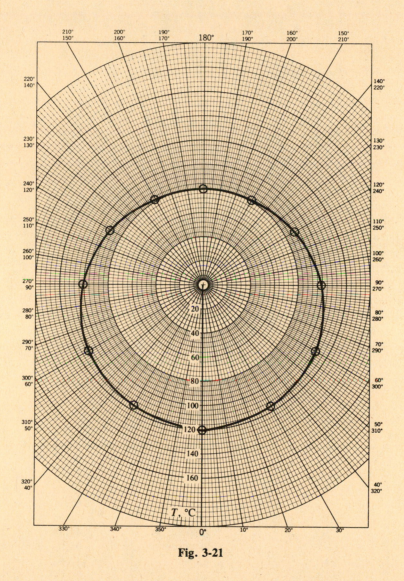

Fig. 3-21

Triangular Coordinates

Triangular coordinates are sometimes used to represent the properties of 3-component systems.   It is convenient to represent such information on triangular coordinate graph paper, as shown in Fig. 3-22.   Each of the vertices represents a single component.   Two-component mixtures are plotted along the edges (axes) of the graph, whereas 3-component mixtures are plotted within the interior.   The use of such graphs depends upon the geometrical fact that the distances (coordinates) of an interior point from the three sides of an equilateral triangle have a fixed sum (e.g. 100%).

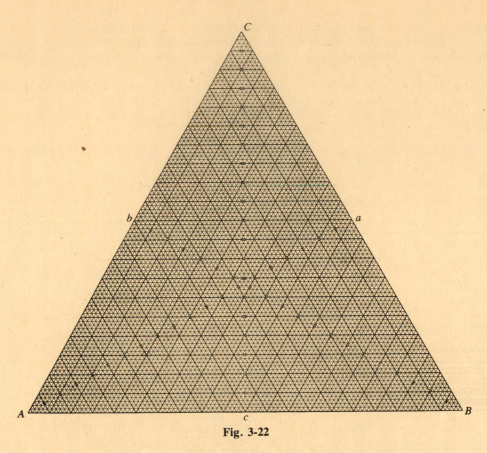

Fig. 3-22

Example 3.17. Benzene and acetone are common organic chemicals that are used to synthesize many different plastics, textiles, and pharmaceuticals.   If benzene, acetone, and water are mixed together in certain proportions, two distinct liquid phases will be formed.   The particular compositions that will separate in this manner can be represented as a special region on a triangular graph.   Several points that lie on the boundary of this region are listed in Table 3-6.

Table 3-6

Benzene, %	Acetone, %	Water, %
0.1	0.0	99.9
0.1	10.0	89.9
0.3	20.0	79.7
0.7	30.0	69.3
1.4	40.0	58.6
3.2	50.0	46.8
9.0	60.0	31.0
24.1	65.3	10.6
99.9	0.0	0.1
89.8	10.0	0.2
79.6	20.0	0.4
69.3	30.0	0.7
58.5	40.0	1.5
47.1	50.0	2.9
34.2	60.0	5.8

The data are shown plotted on a triangular graph in Fig. 3-23.   Notice that a smooth curve has been drawn through the data.   This curve forms the boundary between the single-phase and the two-phase regions of the graph.

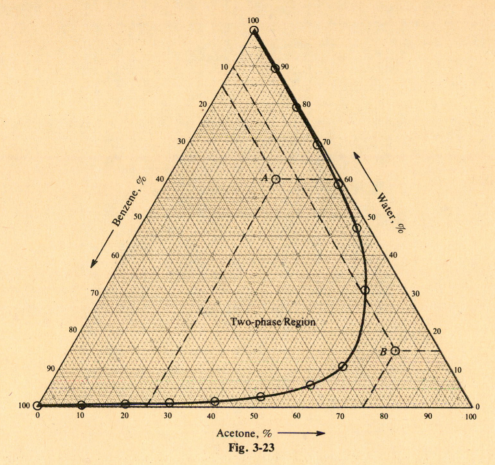

Fig. 3-23

We see that the lower-left vertex in Fig. 3-23 represents pure benzene, the lower-right vertex represents pure acetone and the upper vertex represents pure water.   Each axis represents a mixture of the two components represented by its terminal vertices.   Thus the point labeled "20" on the lower (horizontal) axis represents a mixture of 20 percent acetone and 80 percent benzene.

The interior points in Fig. 3-23 represent various 3-component mixtures of benzene, acetone, and water. Thus point A represents a mixture of 25 percent acetone, 60 percent water, and 15 percent benzene, as indicated by the dashed lines.   (Notice that, for each component, the corresponding dashed line runs parallel to the axis *opposite* the appropriate vertex.)   This point falls within the two-phase region, which is bounded by the curve through the above data points and the benzene–water axis.   Therefore a mixture of 25 percent acetone, 15 percent benzene, and 60 percent water will separate into two distinct phases—one containing mostly water and acetone, the other consisting primarily of benzene and acetone.   The composition of each phase will be represented by a point somewhere on the boundary curve.   The exact location of these points is, however, beyond the scope of this discussion.

In contrast, point B represents a mixture of 75 percent acetone, 15 percent water, and 10 percent benzene. This point does not fall within the two-phase region.   Hence a mixture having this composition will not separate into two distinct phases.

Probability Coordinates

Cumulative distribution data (see Section 3.2) are sometimes plotted on *probability* graph paper.   The ordinate on a probability graph is scaled in such a manner that a cumulative *normal* distribution can be plotted as a straight line.   (The *normal* or *Gaussian* distribution is a symmetric, bell-shaped distribution whose mathematical properties are well known.   The variation in many physical and biological phenomena is governed by the normal distribution.   We will discuss the normal distribution in greater detail in Section 4.8.)   The applicability of the normal distribution to a set of data can therefore be determined by plotting the cumulative relative frequencies of the data on probability paper and observing whether or not a straight line is obtained.   Figure 3-24 shows a typical sheet of probability graph paper.

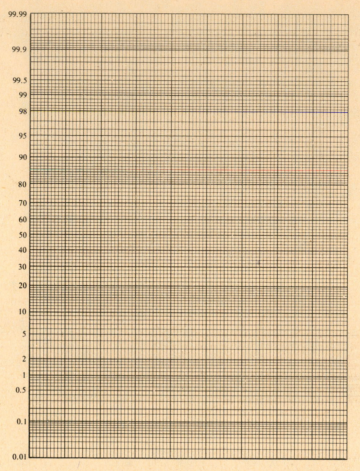

Fig. 3-24

When plotting cumulative distribution data for discrete intervals, it is customary to plot the upper boundary for each interval, rather than the midpoint, along the abscissa.

Example 3.18. A factory manufactures integrated circuits ("chips") for electronic calculators. The industrial engineering department has obtained data on the number of defective chips manufactured during each of several 8-hour shifts; these data are shown in the first two columns of Table 3-7. The two remaining columns were derived from this information, using the method described in Section 3.2.

Table 3-7

Defects per Shift	Number of Occurrences	Relative Frequency, %	Cumulative Relative Frequency, %
0–2	0		
3–4	2	3.6	3.6
5–6	4	7.3	10.9
7–8	8	14.5	25.4
9–10	12	21.8	47.2
11–12	13	23.6	70.8
13–14	9	16.4	87.2
15–16	5	9.1	96.3
17–18	2	3.6	99.9
TOTALS	55	99.9%	

Figure 3-25 shows a probability plot of the cumulative distribution data.   We see that all of the data, except the last data point, fall along a straight line.   Therefore we conclude that the variation in the number of defective chips over several 8-hour shifts is governed by a normal distribution.

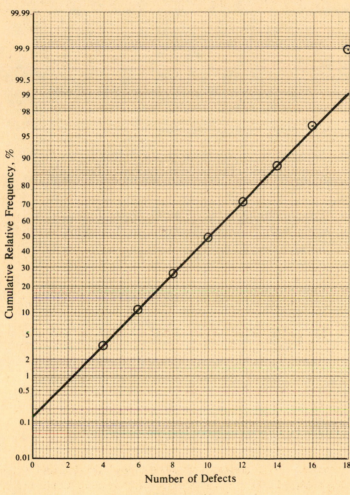

Fig. 3-25

3.6 NOMOGRAPHS

A *nomograph* is a pictorial representation of an algebraic or other equation.   Unlike an ordinary graph, a nomograph does not employ a two-dimensional coordinate system.   Rather, it consists of several scales, usually placed parallel to one another, whose units and relative position are determined by the corresponding equation.   The nomograph is read by associating particular points located along the various scales.

We will consider two types of nomographs in this book: *functional charts* and *alignment charts*.

Functional Charts

A functional chart is used to represent an equation of the form $y = ax + b$,   where a and b are known constants ($b = 0$ in many cases of practical interest).   This type of nomograph consists of two scales that are adjacent to each other.   A functional chart is read by locating a point on one of the scales and determining the value corresponding to that point from the other scale.

Example 3.19. Figure 3-26 shows a functional chart for converting between feet and meters. This functional chart is based upon the equation $F = 3.28M$, where F represents feet and M represents meters (see Appendix A).

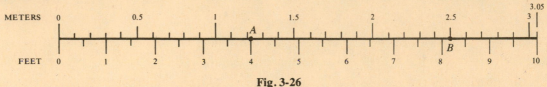

Fig. 3-26

From Fig. 3-26 we see that a value of 4 feet corresponds to approximately 1.22 meters (see point A). Similarly, a value of 2.5 meters is equivalent to approximately 8.2 feet (see point B). Additional equivalences can be established by selecting other points along the chart.

It is quite easy to construct a functional chart for the equation

$$y = ax + b \qquad (a > 0)$$

The lower scale is first divided into equal intervals, where each interval represents a unit of x (or a fixed multiple thereof). The left boundary represents x_{min}, the lowest value of x (often chosen as 0), and the right boundary represents x_{max}, the highest value of x.

After the lower scale has been drawn, the upper scale is divided into equal intervals, with each interval $1/a$ times as large as an interval on the lower scale. Each of the upper intervals will now represent a unit of y (or a fixed multiple thereof). The left boundary will represent y_{min}, and the right boundary y_{max}, where

$$y_{min} = ax_{min} + b \qquad y_{max} = ax_{max} + b$$

Once the upper and lower scales have drawn and labeled, each scale can be further divided into convenient subintervals if desired. A simple modification of the above construction handles the case $a < 0$ (see the discussion of alignment charts, below).

Example 3.20. Construct a functional chart of the equation

$$3y = 2x + 6$$

for the interval $-3 \leqslant x < 9$.

We first divide by 3 in order to put the equation into the form $y = ax + b$. Thus

$$y = \frac{2}{3}x + 2$$

so that $a = 2/3$ and $b = 2$.

The lower scale of the functional chart can now be constructed, as shown in Fig. 3-27(a). We see that the scale is divided into intervals of one x-unit, ranging from $x_{min} = -3$ to $x_{max} = 9$.

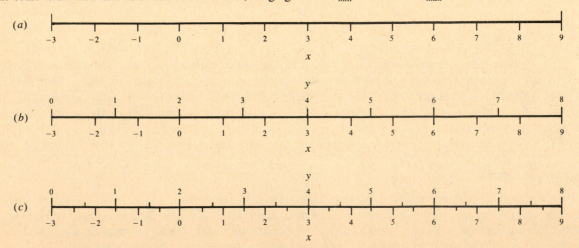

Fig. 3-27

The upper scale is added in Fig. 3-27(b). Each interval along the upper scale is $1/a = 1.5$ times as large as each interval along the lower scale. Notice that

$$y_{\min} = \frac{2}{3}\,(-3) + 2 = 0 \qquad y_{\max} = \frac{2}{3}\,(9) + 2 = 8$$

Finally, Fig. 3-27(c) shows the completed functional chart, after the upper and lower scales have been further divided into convenient subintervals.

Alignment Charts

An alignment chart is used to represent an equation that contains three or more variables. A separate scale must be constructed for each variable. The alignment chart is read by locating a point on each of two of the scales (thus selecting values for two of the variables) and connecting these two points with a straight line. This line is then extended, if necessary, until it intersects a third scale. The point of intersection determines the value of the third (unknown) variable.

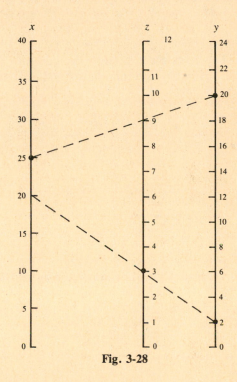

Fig. 3-28

Example 3.21. Figure 3-28 shows an alignment chart for an equation which relates the variables x, y, and z. The chart consists of three parallel scales, each of which is divided into equal intervals. Notice, however, that the interval size and the number of units per interval vary from one scale to another.

Let us make use of Fig. 3-28 to determine the value of z corresponding to $x = 25$ and $y = 20$. Also, let us find the value of x corresponding to $y = 2$ and $z = 3$.

To solve the first part of the problem we locate the points $x = 25$, $y = 20$ and connect them with a straight line, as shown in Fig. 3-28. This line intersects the z-scale at $z = 9$. Hence $z = 9$ when $x = 25$ and $y = 20$.

The second part of the problem is solved by locating the points $y = 2$, $z = 3$ and connecting them with a straight line (see Fig. 3-28). This line is then extended to the x-scale, where it intersects at the point $x = 20$. Thus $x = 20$ when $y = 2$ and $z = 3$.

We now turn our attention to the construction of an alignment chart for the equation

$$z = ax + by \qquad (a > 0, \quad b > 0)$$

We begin by drawing a horizontal line, and then erecting perpendiculars at each end. These perpendiculars will represent the x-scale and the y-scale. The length of the perpendiculars, as well as the distance between them, is arbitrary.

Each perpendicular is then divided into equal intervals. Each interval along the x-scale (the left perpendicular) will correspond to a unit of x (or a fixed multiple thereof), and each unit along the y-scale (the right perpendicular) will represent a unit of y (or a fixed multiple thereof).

Suppose that p represents the length of one x-unit, and q represents the length of one y-unit. The z-scale can now be located between the x-scale and the y-scale. To do so, we erect a third perpendicular between the x- and y-scales, such that

$$\frac{L_1}{L_1 + L_2} = \frac{bp}{bp + aq}$$

where L_1 and L_2 are the distances shown in Fig. 3-29.

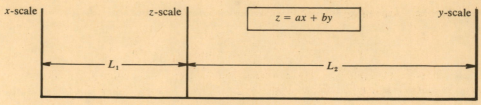

Fig. 3-29

Once the location of the z-scale has been established and a perpendicular has been drawn, the length of one z-unit, r, is calculated from the equation

$$r = \frac{pq}{bp + aq}$$

The z-scale can then be divided into equal intervals of length r (or some multiple of r).

Finally, the minimum z-value is determined from

$$z_{\min} = ax_{\min} + by_{\min}$$

It is then a simple matter to label the z-scale, thus completing the alignment chart.

Example 3.22. Construct an alignment chart of the equation

$$2z = 12x + 4y$$

for the interval $-6 \leqslant x \leqslant 6$ and $0 \leqslant y \leqslant 36$.

We begin by rearranging the equation into the form $z = ax + by$.   Therefore we divide by 2, resulting in

$$z = 6x + 2y$$

so that $a = 6$ and $b = 2$.

Now let us draw a horizontal line 16 cm long, with a perpendicular at each end, as illustrated in Fig. 3-30(a). (Note that the figure has been reduced in size.) The left perpendicular will represent the x-scale and the right perpendicular will represent the y-scale. Each perpendicular is divided into intervals of 2 cm. Notice, however, that an interval on the x-scale represents one x-unit, whereas an interval on the y-scale represents three y-units. Thus $p = 2$ and $q = 2/3$.

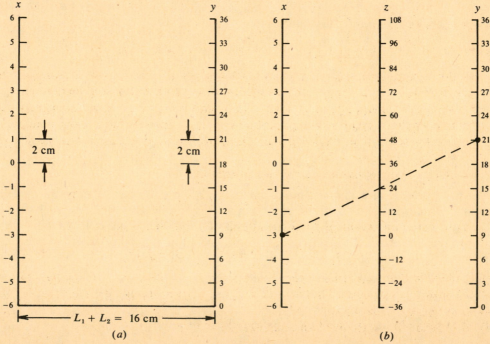

Fig. 3-30

We now locate the z-scale using the equation

$$\frac{L_1}{L_1 + L_2} = \frac{bp}{bp + aq} = \frac{(2)(2)}{(2)(2) + (6)(2/3)} = \frac{4}{4 + 4} = \frac{1}{2}$$

Thus the z-scale will be located midway between the x- and y-scales, as shown in Fig. 3-30(b).

The length of each z-unit can be calculated as

$$r = \frac{pq}{bp + aq} = \frac{(2)(2/3)}{(2)(2) + (6)(2/3)} = \frac{4/3}{4 + 4} = \frac{1}{6}$$

Therefore each z-unit will correspond to (1/6) cm.   If we divide the z-scale into 2 cm intervals, then each interval will represent 12 units.

In order to establish the minimum z-value, we write

$$z_{\min} = ax_{\min} + by_{\min} = (6)(-6) + (2)(0) = -36$$

The z-scale can now be labeled as in Fig. 3-30(b).

Let us check the accuracy of the alignment chart by comparing a calculated value of z with a value obtained from Fig. 3-30(b).   Thus, if $x = -3$ and $y = 21$, the given equation yields

$$z = (6)(-3) + (2)(21) = -18 + 42 = 24$$

The dashed line in Fig. 3-30(b) also indicates that $z = 24$ when $x = -3$ and $y = 21$, as expected.

The method is modified somewhat if one of the terms has a negative coefficient.   The scales are constructed in the same manner, ignoring the minus sign associated with the negative coefficient.   However, the scale corresponding to the negative coefficient is labeled in the opposite direction.   The determination of the minimum z-value is also changed.   If, for example, $a > 0$ but $b < 0$, then

$$z_{\min} = ax_{\min} + by_{\max}$$

(See Problem 3.6.)

If an equation contains four or more variables it is still possible to construct an alignment chart, in much the same manner as outlined above.   Consider, for example, the equation

$$w = ax + by + cz$$

This equation can be written as

$$w = v + cz \qquad \text{where} \qquad v = ax + by$$

Therefore, we first construct an alignment chart for the equation $v = ax + by$.   A second alignment chart, for the equation $w = v + cz$, is then superimposed over the first alignment chart.

In order to read this type of alignment chart, the x- and y-scales are first connected by a straight line which intersects the v-scale at some particular point.   This point is then connected to the z-scale by a second straight line.   The intersection of this second line with the w-scale yields the desired solution.   (See Problem 3.7.)

A linear equation connecting three arbitrary functions,

$$f_1(z) = af_2(x) + bf_3(y)$$

can also be represented by means of an alignment chart.   To do so, let

$$w = f_1(z) \qquad u = f_2(x) \qquad v = f_3(y)$$

An alignment chart is then constructed for the equation $w = au + bv$.   Once the chart has been completed, the scales can be labeled in terms of x, y, and z, even though the scales were constructed in terms of u, v, and w.   (See Problem 3.8.)

Finally, it is possible to represent an equation of the form

$$z = cx^a y^b$$

by means of an alignment chart. This is accomplished by taking logarithms of both sides of the equation, i.e.

$$\log z = \log c + a \log x + b \log y$$

Now let $w = z/c$, so that

$$\log w = \log z - \log c = a \log x + b \log y$$

An alignment chart involving $\log w$, $\log x$, and $\log y$ can now be constructed in the customary manner. It is usually desirable to label the scales in terms of x, y, and z, even though they were constructed in terms of $\log x$, $\log y$, and $\log w$. The procedure is illustrated in Problem 3.9.

Solved Problems

3.1. The following data describe the shear force (F) in a beam as a function of position (x). Plot the data on a cartesian graph and determine an equation that can be used to represent F as a function of x.

x, m	0	1	2	3	4	5	6	7	8	9	10
F, kN	−1.67	−1.00	−0.33	0.33	1.00	1.67	2.33	3.00	3.67	4.33	5.00

The data are plotted in Fig. 3-31. A straight line can be passed through the data, having slope 0.67 and vertical intercept −1.67. Thus

$$F = 0.67x - 1.67 \qquad \text{or} \qquad 3F = 2x - 5$$

Note that F represents kilonewtons (thousands of newtons).

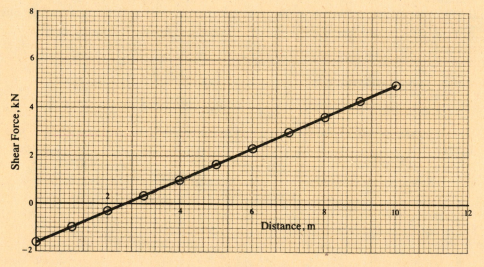

Fig. 3-31

3.2. The data below refer to the growth of a bacteria culture. The population density (d), in parts per million, is tabulated as a function of time (t). Plot the data on a semilog graph and determine an equation that will represent the population density as a function of time.

t, days	0	0.5	1.0	1.5	2.0	2.5
d, ppm	0.10	0.45	2.01	9.00	40.34	180.80

The first 5 points can be plotted on 3-cycle semilog paper, as shown in Fig. 3-32.   A straight line can be passed through the data.   The slope of this line is

$$\frac{\log 2.01 - \log 0.10}{1 - 0} = \frac{0.30 - (-1.00)}{1} = 1.30$$

and the vertical intercept is $\log 0.10$.   We know that the equation $d = ae^{bt}$ can be represented as a straight line on semilog paper.   Therefore we can write

$$\ln d = \ln a + bt \qquad \text{or} \qquad \log d = \log a + 0.434\,bt$$

since $\log x = 0.434 \ln x$.   We have found that the vertical intercept is $\log a = \log 0.10$ and the slope is

$$0.434\,b = 1.30 \qquad \text{or} \qquad b = 3.0$$

Therefore $d = 0.10e^{3.0t}$.

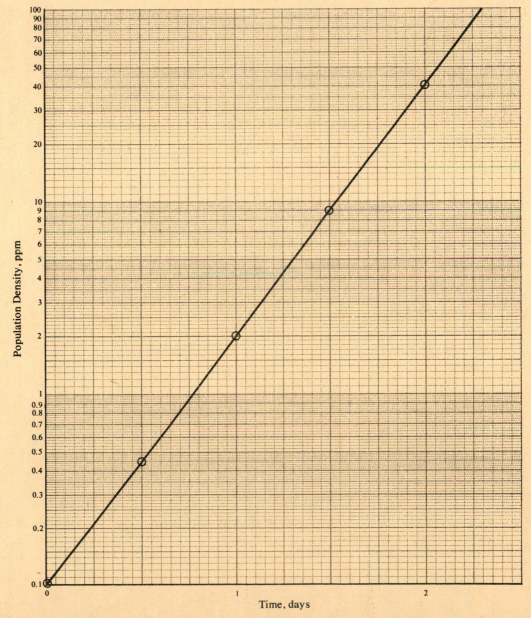

Population Density, ppm

Time, days

Fig. 3-32

3.3.   When a fluid is flowing in a pipe, the fluid velocity is greater at the centerline of the pipe than at the pipe wall.   The following data describe fluid velocity (v) as a function of radial distance from the pipe wall (r).   Plot the data on a log-log graph and determine an equation that will represent velocity as a function of distance from the wall.

r, cm	0	2	4	6	8	10
v, cm/s	0	2.6	4.6	6.3	7.9	9.5

The last five points can be plotted on one-cycle log-log paper, as shown in Fig. 3-33.   A straight line can be passed through the data.   The slope of this line is

$$\frac{\log 9.5 - \log 1.5}{\log 10 - \log 1} = \frac{0.978 - 0.176}{1 - 0} = 0.802$$

We know that the equation $v = ar^b$ can be represented as a straight line on log-log paper.   Thus

$$\log v = \log a + b \log r = \log a + 0.802 \log r$$

Also, we see that $v = 1.5$ cm/s   when $r = 1$ cm.   Thus

$$\log 1.5 = \log a + 0.802 \log 1 = \log a$$

Therefore $a = 1.5$ and the desired equation is

$$v = 1.5 r^{0.802}$$

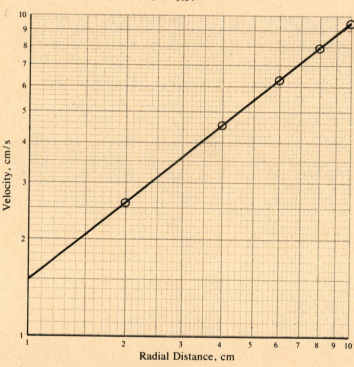

Fig. 3-33

3.4.   An instrument manufacturer has obtained the following data for the number of defective thermometers manufactured each day over a period of 200 days.

Defects per Day	0	1	2	3	4	5	6	7	8	9	10
Number of Days	2	5	14	24	35	40	36	25	13	4	2

Plot the cumulative relative distribution on arithmetic graph paper and on probability paper. Can anything be inferred about the nature of the distribution from either of these plots?

In order to plot the cumulative distribution Table 3-8 is first constructed.

Table 3-8

Number of Defects per Day	Number of Days	Cumulative Number of Days	Cumulative Relative Frequency, %
0	2	2	1.0
1	5	7	3.5
2	14	21	10.5
3	24	45	22.5
4	35	80	40.0
5	40	120	60.0
6	36	156	78.0
7	25	181	90.5
8	13	194	97.0
9	4	198	99.0
10	2	200	100.0
TOTAL	200		

Figure 3-34 shows a plot of the cumulative distribution on rectangular coordinates, in the form of a bar graph. The same data are shown plotted on probability paper in Fig. 3-35. A straight line can be passed through all of the data in Fig. 3-35. We therefore conclude that the number of defects per day is normally distributed.

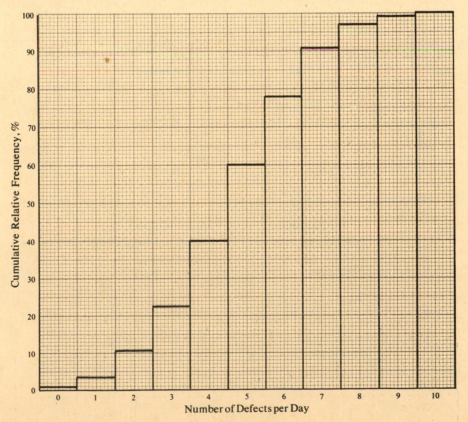

Fig. 3-34

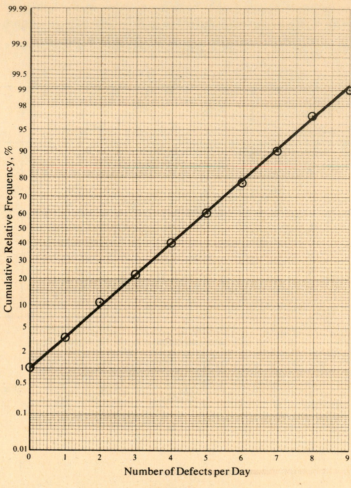

Fig. 3-35

3.5. The following data describe the gain, in decibels, of a symmetric, highly directional television antenna.

Angle, deg	0	30	60	90	120	150	180
Gain, dB	42	40	36	31	25	19	12

Plot the data on cartesian coordinates and on polar coordinates for the entire 360° range of the antenna.

Figure 3-36 shows the data plotted on cartesian coordinates. It was possible to cover the entire 360° range because of the symmetric nature of the antenna. (Notice the symmetry about the vertical 180° line in Fig. 3-36.) The same data are plotted on polar coordinates in Fig. 3-37.

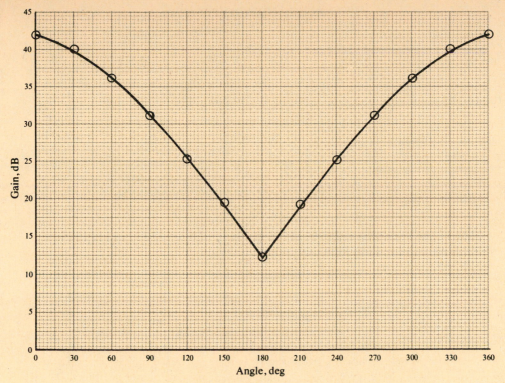

Fig. 3-36

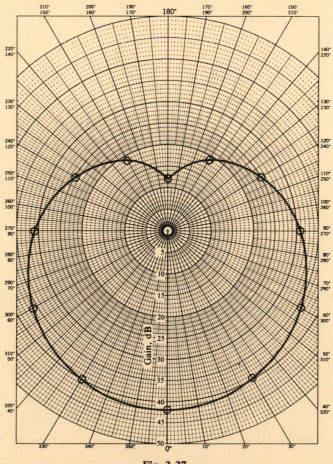

Fig. 3-37

3.6. Construct an alignment chart for the equation

$$z = 6x - 2y$$

for the interval $-6 \leqslant x \leqslant 6$ and $0 \leqslant y \leqslant 36$.

This equation is identical to that presented in Example 3.22, except for the negative coefficient of y. Thus the layout of the desired alignment chart will be identical to the layout of Fig. 3-30(b), i.e.

$$p = 2 \qquad q = \frac{2}{3} \qquad \frac{L_1}{L_1 + L_2} = \frac{1}{2} \qquad r = \frac{1}{6}$$

Because of the negative coefficient, however, the y-scale will be labeled in the reverse direction; that is, the values will *decrease*, from $y = 36$ to $y = 0$, as we move up the y-scale. Furthermore, since

$$z_{min} = 6x_{min} - 2y_{max}$$
$$= (6)(-6) - (2)(36) = -108$$

the z-scale will now range from $z = -108$ to $z = 36$. The completed alignment chart is shown in Fig. 3-38.

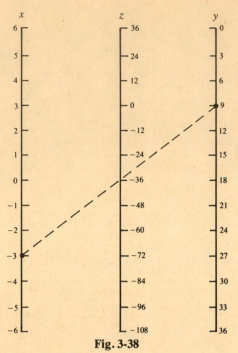

Fig. 3-38

To check the accuracy of the chart, let us determine the value of z corresponding to $x = -3$ and $y = 9$. The alignment chart shows a value $z = -36$, as indicated by the dashed line in Fig. 3-38. If we substitute the given values into the original equation, we obtain

$$z = (6)(-3) - (2)(9) = -18 - 18 = -36$$

which confirms the value of z read from the alignment chart.

3.7. Construct an alignment chart for the equation

$$w = 4x + y + 2z$$

for $0 \leqslant x \leqslant 25$, $0 \leqslant y \leqslant 100$, $0 \leqslant z \leqslant 50$.

We first construct an alignment chart for $v = 4x + y$. To do so, construct two vertical lines 10 cm (or some other convenient distance) apart. These lines will become the x- and y-scales. Let us divide each scale into 1 cm intervals, where each interval on the x-scale will represent 2 units and each y-interval will represent 8 units. Thus $p = 1/2$ and $q = 1/8$. We now locate the v-scale, as

$$L_1 = (L_1 + L_2)\left(\frac{bp}{bp + aq}\right) = (10)\left[\frac{(1)(1/2)}{(1)(1/2) + (4)(1/8)}\right] = 5 \text{ cm}$$

Hence the v-scale is located midway between the x-scale and the y-scale, as shown in Fig. 3-39(a). The length of each v-unit will be

$$r = \frac{pq}{bp + aq} = \frac{(1/2)(1/8)}{(1)(1/2) + (4)(1/8)} = \frac{1}{16}$$

If we divide the v-scale into 1 cm intervals, then each interval will represent 16 units. Figure 3-39(a) shows the proper layout of the x-, y-, and v-scales.

We now construct another alignment chart, for the equation $w = v + 2z$, and superimpose this alignment chart over the previous chart. Let us locate the z-scale 9 cm to the right of the v-scale, and let us divide the z-scale into 1 cm intervals, where each interval represents 4 units. Thus we have a new set of constants:

$$a = 1 \qquad b = 2 \qquad L_1 + L_2 = 9 \qquad p = \frac{1}{16} \qquad q = \frac{1}{4}$$

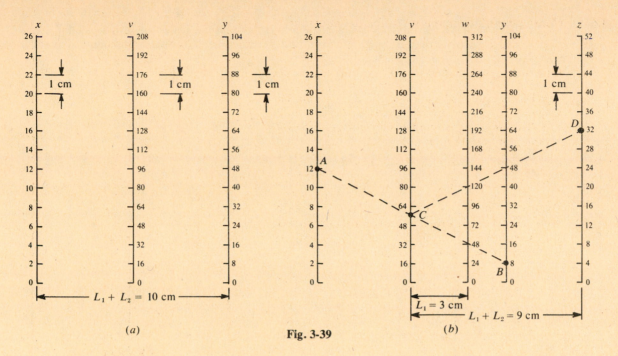

Fig. 3-39

The w-scale can now be located as

$$L_1 = (L_1 + L_2)\left(\frac{bp}{bp + aq}\right) = (9)\left[\frac{(2)(1/16)}{(2)(1/16) + (1)(1/4)}\right] = 3 \text{ cm}$$

so that the w-scale is placed 3 cm to the right of the v-scale, as shown in Fig. 3-39(b). The length of each w-unit will be

$$r = \frac{pq}{bp + aq} = \frac{(1/16)(1/4)}{(2)(1/16) + (1)(1/4)} = \frac{1}{24}$$

If the w-scale is divided into 1 cm intervals, then each interval will represent 24 units. Figure 3-39(b) shows the complete alignment chart.

To see how the alignment chart is used, let us determine the value of w corresponding to the values $x = 12$, $y = 8$, and $z = 32$. We first connect the points $x = 12$ and $y = 8$ with a straight line [line AB in Fig. 3-39(b)], which intersects the v-scale at $v = 48$ (point C). We then connect this point with the point $z = 32$ (line CD). This second line intersects the w-scale at $w = 120$. Thus $w = 120$ when $x = 12$, $y = 8$, and $z = 32$. We can easily verify the accuracy of this result by substituting into the original equation, i.e.

$$w = (4)(12) + 8 + (2)(32) = 120$$

It was actually not necessary for us to have labeled the v-scale, since its only purpose is to provide an end point for the second line [line CD in Fig. 3-39(b)]. The numerical value corresponding to point C is irrelevant.

3.8. Construct an alignment chart for the equation

$$z = 2x^{0.5} + 3y^{0.7}$$

for $0 \leq x \leq 25$ and $0 \leq y \leq 10$.

Let $u = x^{0.5}$ and $v = y^{0.7}$. Thus the given equation becomes

$$z = 2u + 3v$$

for $0 \leq u \leq 5$ and $0 \leq v \leq 5$. (Note that $\log 5 \approx 0.7$.)

The construction of an alignment chart for z in terms of u and v is straightforward; Fig. 3-40(a) shows the result which is obtained.

In Fig. 3-40(b) we see the same alignment chart, except that the scales are now labeled in terms of x, y, and z rather than u, v, and z.   The x- and y-scales were constructed by determining, for example, that $x = 3$ corresponds to

$$u = 3^{0.5} = 1.73$$

$x = 5$ corresponds to $u = 2.24$,   and so on.

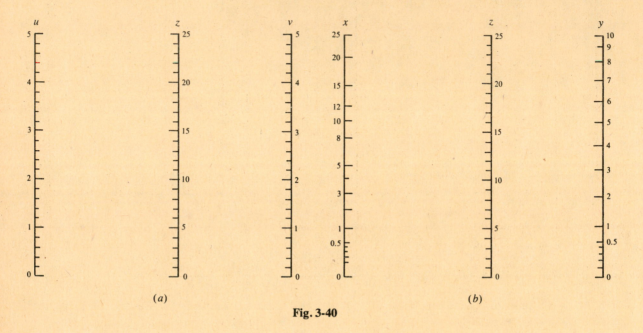

(a)   (b)

Fig. 3-40

3.9.   The sulfur content of coal is often expressed in terms of the number of pounds of sulfur dioxide (SO_2) released per million Btu when the coal is burned.   Mathematically, this can be written

$$Y = \frac{20S}{E}$$

where Y = lbs SO_2 per million Btu
            S = weight percent sulfur in the coal
            E = energy content of the coal, in thousands of Btu per pound

Construct an alignment chart for this equation over $0.3 \leqslant S \leqslant 10$, $8 \leqslant E \leqslant 16$.

Taking the logs of both sides, we obtain

$$\log Y = \log (20S) - \log E \qquad \text{or} \qquad w = u - v$$

where $w = \log Y$
          $u = \log (20S)$ $0.7 \leqslant u \leqslant 2.3$
          $v = \log E$ $0.9 \leqslant v \leqslant 1.2$

The construction of an alignment chart in terms of u, v, and w is straightforward.   (Note that the v-scale will be labeled in the reverse direction because of the negative coefficient.)   Figure 3-41(a) shows the result which is obtained.

It is now a simple matter to relabel the scales in terms of S, E, and Y.   (Note, for example, that $S = 1$ corresponds to

$$u = \log 20 = 1.3$$

$S = 2$ corresponds to $u = 1.6$,   and so on.)   Figure 3-41(b) shows the desired alignment chart, labeled in terms of S, E, and Y.

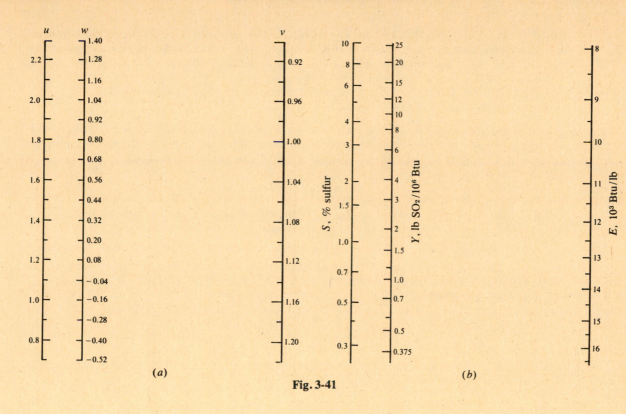

(a)

(b)

Fig. 3-41

Supplementary Problems

Answers to most problems are provided at the end of the book.

3.10.   Figure 3-1 shows the height of a falling object as a function of time.   Use Fig. 3-1 to answer the following questions.   (a)   How high will the object be after 2 s?   (b)   How high will the object be after 5 s?   (c)   How far will the object have fallen between the end of the first and the end of the third second?   (d)   How long will it take the object to reach a height of 100 m?   (e)   How much time is required for the object to fall the last 50 m?   (f)   How long will it take the object to fall from a height of 140 m to a height of 70 m?   (g)   If the object were initially 150 m above the ground, how long would it take to fall all the way to the ground?

3.11.   Figure 3-2 is a plot of concentration of a chemical substance versus time.   Use this figure to answer the following questions.   (a)   What is the concentration after 4.2 s?   (b)   Approximately what will the concentration be after 10 s?   (c)   How long will it take for the concentration to reach the value 1.1 mol/L?   (d)   How long will it take for the concentration to increase from 1 to 2 mol/L?   (e)   What is the difference (error) between the actual (measured) concentration and the value read from the curve at 3 s?   What is the corresponding percent error?

3.12.   Figure 3-3 shows the results of three different tests which measure cumulative probability of failure as a function of temperature.   Each test corresponds to a different initial (critical) temperature.   Use Fig. 3-3 to answer the following questions.   (a)   What is the cumulative probability of failure for test A, test B, and test C at 1295 K?   (b)   What is the median temperature (i.e. the temperature corresponding to the 50% cumulative probability of failure) for each test?   (c)   Suppose that a test had been conducted with a critical temperature of 1235 K.   What would be the median temperature for this test?   At what temperature would the cumulative probability of failure reach 100%?   (d)   Suppose that a test had been conducted with a critical temperature of 1270 K.   What would be the median temperature for this test?   What would be the cumulative probability of failure at 1320 K?

3.13.   Use Fig. 3-4 to answer the following questions.   (a)   What change in force will result if the displacement is increased from 1 to 3 cm?   (b)   What change in force will result if the displacement is increased from −1 to +1 cm?   (c)   What is the significance of the sign in the above answers?

3.14.   Use Fig. 3-5 to answer the following questions.   (a)   What value of stress corresponds to a strain value of 0.0018?   (b)   What value of strain corresponds to a stress value of 450×10^7 N/m^2?   (c)   What value of strain corresponds to a stress value of 450×10^{-5} N/m^2?

3.15.   Determine the answers to the following questions from Fig. 3-7.   (a)   What relative frequency corresponds to the range of diameters 89.5–89.7 mm?   (b)   What range of diameters corresponds to a relative frequency of 53.5%?

3.16.   Pass a smooth curve through the right interval boundaries in Fig. 3-8 and then answer the following questions.   (a)   What cumulative relative frequency corresponds to a diameter of 89.65 mm?   (b)   What diameter corresponds to a cumulative relative frequency of 53.5%?   (c)   What diameter corresponds to a cumulative relative frequency of 10%?

3.17.   Use Fig. 3-12 to answer the following questions.   (a)   What value of y corresponds to $x = 5.5$?   (b)   What value of y corresponds to $x = 9.2$?   (c)   What value of x corresponds to $y = 4$?   (d)   What value of x corresponds to $y = 0.4$?   (e)   What is the slope of the line?

3.18.   Use Fig. 3-13 to answer the following questions.   (a)   What value of k corresponds to $T = 300$ K?   (b)   What temperature corresponds to $k = 400 \times 10^5$ s^{-1}?   (c)   What change in temperature is required for k to decrease from 100×10^5 s^{-1} to 10×10^5 s^{-1}?   (d)   What is the slope of the line?

3.19.   Answer the following questions with the aid of Fig. 3-17.   (a)   What value of y corresponds to $x = 4.5$?   (b)   What value of y corresponds to $x = 2.5$?   (c)   What value of x corresponds to $y = 0.3$?   (d)   What value of x corresponds to $y = 3$?   (e)   What is the slope of the line?

3.20.   Determine the answers to the following questions by use of Fig. 3-21.   (a)   What is the temperature at $\theta = 45°$?   135°?   315°?   −135°?   (b)   What is the temperature at $\theta = 15°$?   80°?   110°?   320°?   (c)   At what angular positions will the temperature be 95 °C?

3.21.   The composition of a mixture of component X, component Y, and component Z can be determined from Fig. 3-42.   What is the composition at   (a)   point A?   (b)   point B?   (c)   point C?   (d)   point D?   (e)   point E?

3.22.   Use Fig. 3-23 to answer the following questions.   (a)   How much benzene and acetone are present on the boundary separating the single-phase and the two-phase regions if the water concentration is 20%?   (b)   Will the following point lie in the single-phase or the two-phase region: 25% water, 20% acetone, 55% benzene?   (c)   Will the following point lie in the single-phase or the two-phase region: 25% water, 65% acetone, 10% benzene?   (d)   If the points in (b) and (c) are connected by a straight line, at what composition will the line intersect the boundary between the single-phase and the two-phase regions?

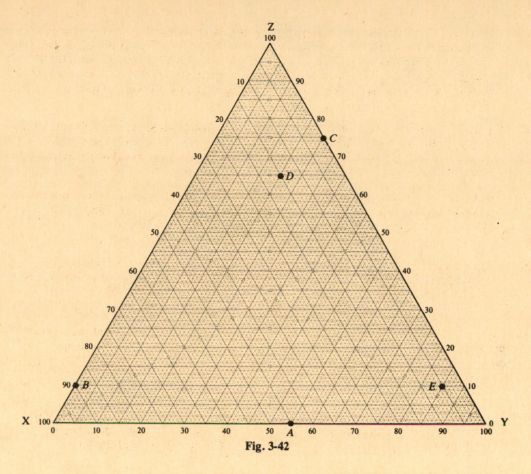

Fig. 3-42

3.23. From Fig. 3-25, what change in the cumulative distribution corresponds to an increase in the number of defects from (a) 10 to 12? (b) 2 to 4?

3.24. The following data describe the force exerted by a spring as it is stretched from its equilibrium position. Plot the data on an arithmetic graph and determine an equation that will represent force as a function of displacement.

Displacement, m	−4	−2	0	2	4
Force, N	120	60	0	−60	−120

3.25. Suppose that the spring described in Problem 3.24 were characterized by the following data. Plot the data and determine how the force can be represented mathematically as a function of displacement.

Displacement, m	−5	−4	−3	−2	−1	0	1	2	3	4	5
Force, N	90	72	54	36	18	0	−12	−24	−36	−48	−60

3.26. The following data indicate the level of an atmospheric pollutant, in parts per million (ppm), at various times during a given 24-hour period. Plot the data on an arithmetic graph. Estimate the maximum and minimum pollution levels, and the time of day when they occur.

Time, h:min	0:00	2:00	4:00	6:00	7:00	8:00	9:00	10:00	11:00	12:00
Air Quality, ppm	12.2	12.5	11.9	13.5	22.4	31.4	57.7	84.4	68.0	51.6
Time	13:00	14:00	15:00	16:00	17:00	18:00	19:00	20:00	22:00	24:00
Air Quality	46.6	43.1	39.2	44.7	62.9	88.0	71.4	59.0	43.5	28.7

3.27. Prepare a bar graph for the data presented in Problem 3.26. Let 12.5 ppm represent the air quality for the time interval 1:00–3:00, etc. Connect the midpoints of the intervals with straight-line segments. Compare with the smooth curve obtained in Problem 3.26.

3.28. A photovoltaic device is suddenly exposed to an intense beam of light. The following data describe the voltage drop across the device as a function of time.

Time, ms	0	0.25	0.5	1	2	4	8	16
Voltage, mV	0	0.01	0.02	0.037	0.060	0.084	0.097	0.0999

(a) Plot the data on an arithmetic graph. (b) Transform the data in such a manner that a curve of exponential form can be passed through the plotted points on an arithmetic graph. (*Hint*: Plot $1 - 10V$ against t.) (c) Plot the data on a semilog graph, presenting the data in the same manner as in (b) above. (d) Determine an equation that will accurately express voltage as a function of time.

3.29. The viscosity of a polymeric material varies with temperature in the manner indicated by the following data:

Temperature, K	280	300	320	340	360	380
Viscosity, cP	235	210	190	175	160	150

(a) Plot the data on an arithmetic graph. (b) Plot the data on a log-log graph. (c) Determine an equation that will represent viscosity as a function of temperature over the given temperature range.

3.30. The following data describe the temperature dependence of k, a chemical reaction rate constant (see Example 3.11). Determine a value for A, the frequency factor, and for E, the molar activation energy, in the rate equation

$$k = Ae^{-E/RT}$$

where $R = 8.32 \text{ J/mol} \cdot \text{K}$.

T, K	340	360	380	400	420	440	460	480
k, s^{-1}	3	15	60	220	685	1940	5010	12 000

3.31. The concentration of a radioactive substance is given by the equation

$$y = \frac{c_1}{c_1 - c_2}(e^{-c_2 t} - e^{-c_1 t})$$

(See Example 1.27.). The following data indicate the concentration of this substance at various times.

Time, h	0	1	2	3	4	6	8	10	12	15	20
Concentration	0	0.094	0.178	0.253	0.319	0.430	0.515	0.580	0.628	0.677	0.712
Time	30	40	50	60	80	100	120	140	160	180	200
Concentration	0.692	0.624	0.548	0.475	0.354	0.262	0.194	0.144	0.107	0.079	0.059

Determine the values of c_1 and c_2 by plotting the data on semilog graph paper and passing straight lines through the initial and the final few data points. (*Hint*: If $c_1 > c_2$, then the given equation can be approximated as

$$y = \frac{c_1}{c_1 - c_2} e^{-c_2 t}$$

when t is large. Moreover, when t is close to zero,

$$y = c_1 t$$

approximately.

3.32. A metal rod at 1300 °C is plunged into a bath of oil whose temperature is maintained at 100 °C. The temperature at the center of the rod is given by the expression

$$T = 100 + 1200 e^{-0.1 t}$$

where t is the time (in seconds) that the rod remains in the constant-temperature bath. Show how the equation can be represented as a straight line on semilog paper.

3.33. The speed of sound in a gas is given by the formula

$$u = \sqrt{\frac{\gamma RT}{M}}$$

where u = speed of sound, m/s
   γ = 1.40 for most common gases
   R = universal gas constant, 8.32 J/mol $\cdot$ K
   T = gas temperature, K
   M = molecular weight of gas, kg/mol

Prepare a log-log plot of u against T for   (*a*)   hydrogen ($M = 2 \times 10^{-3}$),   (*b*)   nitrogen ($M = 28 \times 10^{-3}$),   (*c*)   oxygen ($M = 32 \times 10^{-3}$),   (*d*)   air ($M = 29 \times 10^{-3}$).   Plot all the curves on the same graph.   Consider the temperature interval $200 \leq T \leq 400$ K.

3.34. Table 3-9 shows the monthly sales of television sets for a given calendar year. (*a*) Prepare a bar graph of the data. (*b*) Express the monthly sales figures as a percentage of total yearly sales. Prepare a bar graph of the percentage data.

Table 3-9

Month	Number of Units Sold	Month	Number of Units Sold
January	1570	July	1288
February	1417	August	1337
March	1886	September	1848
April	2143	October	2216
May	1705	November	2058
June	1420	December	1790

3.35. Table 3-10 shows the average monthly price of common stock for the Zilch Manufacturing Company over a two-year period. (*a*) Prepare a bar graph of the data on semilog paper. (*b*) Plot the individual points on semilog paper, assuming that each stock price corresponds to the center of the interval (middle of the month). Connect the data points with straight-line segments.

Table 3–10

First Year		Second Year	
Month	Stock Price, $	Month	Stock Price, $
Jan.	3.60	Jan	25.38
Feb.	4.79	Feb.	32.20
March	6.35	March	41.65
April	8.23	April	50.75
May	10.79	May	60.41
June	14.22	June	53.10
July	19.80	July	45.15
Aug.	15.62	Aug.	39.07
Sept.	10.60	Sept.	35.98
Oct.	11.27	Oct.	45.44
Nov.	15.53	Nov.	58.57
Dec.	19.16	Dec.	72.50

3.36. A chemical company manufactures a house paint that is supposed to contain 24.2% pigment. Several random samples of the paint, obtained at arbitrary times, contained the following pigment percentages:

24.22	23.80	23.97	24.30	24.37	24.14
24.87	23.95	24.47	24.32	24.41	24.43
24.19	24.25	24.38	24.40	24.27	23.95
24.05	24.22	24.04	24.62	24.41	24.06

(*a*) Group the data into intervals of 0.1 percentage points (i.e. determine the number of samples between 23.8% and 23.9%, 23.9% and 24.0%, etc.). (*b*) Prepare a bar graph of the grouped data. (*c*) Prepare a histogram showing the percentage of all samples (relative frequency) falling into each interval. (*d*) Prepare a bar graph showing the cumulative relative frequencies of the data plotted in part (*c*).

3.37. Table 3-11 is a summary of the final exam scores in a freshman engineering class. (*a*) Prepare a histogram of the above data. (*b*) Prepare a bar graph showing the cumulative relative frequencies of the above data. (*c*) Plot the cumulative distribution on probability paper. Are the exam scores normally distributed?

Table 3-11

Exam Score Interval	Number of Exams	Exam Score Interval	Number of Exams
96–100	3	56–60	38
91–95	5	51–55	28
86–90	7	46–50	18
81–85	12	41–45	9
76–80	20	36–40	8
71–75	32	31–35	5
66–70	40	26–30	3
61–65	44	21–25	2

3.38. The height of water in a cylindrical tank is related to the discharge velocity by the equation

$$h = 0.65 \log_{10} (1 + 8v)$$

(see Example 1.28). (*a*) Prepare a plot of *h* versus *v* on arithmetic graph paper. Consider the interval $0 \leqslant v \leqslant 0.20$ m/s. (*b*) If we wanted to represent the relationship between *h* and *v* with a straight line, what type of coordinate system would be required?

3.39. The movement of an automatic control device is described by the equation

$$y = 1 - e^{-0.5t}$$

where $y =$ position of controller arm after actuating signal has been received, cm
$t =$ time after actuating signal has been received, s

(*a*) Plot *y* against *t* on arithmetic graph paper, for the interval $0 \leqslant t \leqslant 10$ s. (*b*) Plot *y* against *t*, and $1 - y$ against *t*, on semilogarithmic graph paper. Make both plots on the same graph. Which plot is preferable? (*c*) Plot *y* against *t* on polar graph paper, with *t* as the angular coordinate. Show the direction of increasing *t* on the curve.

3.40. The following data describe the temperature at various angular positions on the circumference of a metal cylinder.

Angle, deg	0	30	60	90	120	150	180
Temperature, °C	50	65	76	80	76	65	50
Angle	210	240	270	300	330	360	
Temperature	35	24	20	24	35	50	

Plot the data on polar coordinate graph paper. Compare with Fig. 3-21 (see Example 3.16). Can you recognize an equation that will represent temperature as a function of angular position?

3.41. Figure 3-26 is a functional chart for converting between feet and meters. Use this chart to answer the following questions. (*a*) Approximately how many meters are equivalent to 5 ft? 9.3 ft? 0.75 ft? 3.25 ft? (*b*) Approximately how many feet are equivalent to 0.5 m? 2.1 m? 2.25 m? 1.67 m?

3.42. Construct a functional chart for converting between ft · lb$_f$ and J.

3.43. Construct a functional chart for converting between lb$_f$ and N.

3.44. Construct a functional chart for converting between lb$_m$ and kg.

3.45. Construct a functional chart for converting between temperature readings in °F and °C. Use the conversion formula

$$\theta_F = 1.8\theta_C + 32$$

Consider the interval $0 \leqslant \theta_F \leqslant 212$.

3.46. Construct a functional chart for converting between gallons and liters.

3.47. Answer the following questions with the assistance of Fig. 3-28. (*a*) What value of *z* corresponds to $x = 30$ and $y = 15$? (*b*) What value of *z* corresponds to $x = 8$ and $y = 12.5$? (*c*) What value of *x* corresponds to $y = 10$ and $z = 6.3$? (*d*) What value of *y* corresponds to $x = 35$ and $z = 8$?

3.48.   Answer the following questions with the aid of Fig. 3-30(b).   (a)   What value of z corresponds to $x = -4.5$ and $y = 0$?   (b)   What value of z corresponds to $x = 4$ and $y = 4$?   (c)   What value of y corresponds to $x = 0.5$ and $z = 12$?   (d)   What value of x corresponds to $y = 33$ and $z = 33$?

3.49.   Use Fig. 3-39(b) to answer the following questions.   (a)   What value of w corresponds to $x = 5$, $y = 45$, and $z = 22$?   (b)   What value of w corresponds to $x = 25$, $y = 0$, and $z = 11$?   (c) What value of z corresponds to $x = 10$, $y = 70$, and $w = 200$?   (d)   What value of x corresponds to $y = 16$, $z = 20$, and $w = 72$?   (e)   What value of y corresponds to $x = 20$, $z = 40$, and $w = 216$?

3.50.   Use Fig. 3-40(b) to answer the following questions.   (a)   What value of z corresponds to $x = 9$ and $y = 2$?   (b)   What value of z corresponds to $x = 0.4$ and $y = 7.5$?   (c)   What value of x corresponds to $y = 4$ and $z = 17$?   (d)   What value of y corresponds to $x = 3.3$ and $z = 17.5$?

3.51.   Figure 3-41(b) shows the sulfur dioxide content per million Btu released when coal having a known energy content and a known sulfur content is burned.   Use this figure to answer the following questions.   (a)   What will be the SO_2 content per million Btu if the coal contains 5% sulfur and its energy content is 12 300 Btu/lb?   (b)   What sulfur content corresponds to 2.5 lb SO_2 per million Btu if the energy content of the coal is 11 500 Btu/lb?   (c)   If a coal has a sulfur content of 3.6% and 9 lb of SO_2 is released per million Btu, what is the energy content of the coal?

3.52.   Construct an alignment chart for the equation

$$4z = 6x + 3y$$

for the interval $0 \leqslant x \leqslant 8$, $0 \leqslant y \leqslant 16$.

3.53.   Construct an alignment chart for the equation

$$3z = 5x - 2y$$

for the interval $0 \leqslant x \leqslant 10$, $0 \leqslant y \leqslant 25$.

3.54.   Construct an alignment chart for the equation

$$3z = 5x - 2y$$

for the interval $5 \leqslant x \leqslant 15$, $-10 \leqslant y \leqslant 15$.

3.55.   The ideal gas law can be written

$$pV = nRT$$

where p = pressure, atm
V = volume of gas, liters
n = number of moles of gas
R = universal gas constant, 0.082 atm $\cdot$ liter/mol $\cdot$ K
T = absolute temperature, K

Prepare an alignment chart of the ideal gas law for the special case where $n = 1$ mol.   Consider the interval $10 \leqslant p \leqslant 100$ atm, $200 \leqslant T \leqslant 500$ K.

3.56.   Construct an alignment chart for the ideal gas law, as described in Problem 3.55, for the interval $1 \leqslant p \leqslant 10$ atm, $100 \leqslant T \leqslant 600$ K, $0.1 \leqslant n \leqslant 20$ mol.

3.57.   The formula $F = (1 + 0.01\,i)^n$ allows us to calculate the compounding of an investment over n years at an annual interest rate i (where i is expressed as a percentage).   Construct an alignment chart for this formula.   Consider the interval $1 \leqslant i \leqslant 15\%$, $1 \leqslant n \leqslant 20$.

Chapter 4

Data Reduction Techniques

Engineers are often faced with the need to gather and interpret measured data. As a rule the raw data will not be particularly informative; however, a great deal of useful information can be extracted from a set of data by utilizing a few simple data reduction procedures.

4.1 MEASURED DATA

The accuracy of a measured value will always be limited by the precision of the measuring device. This limitation must be considered when recording or reporting measured data. The number of significant figures in a measured value should normally be chosen such that uncertainty exists only in the last figure (i.e. the least significant digit).

Example 4.1. The scale on a mercury thermometer is divided into intervals of 0.1 °C. The top of the mercury column is observed either as aligned with a particular division or as located somewhere between two adjacent divisions. Therefore the thermometer has an accuracy of about 0.05 °C.

Suppose that, for a particular measurement, the top of the mercury column is observed to be between the 25.2 °C and the 25.3 °C divisions. If the top of the column appears closer to the 25.3 °C division, then it would be proper to accept 25.30 °C as the correct temperature. Similarly, we would accept 25.20 °C as the correct temperature if the top of the mercury column were closer to this division.

Notice that both the above temperatures are expressed in terms of four significant figures. The first three figures can be read directly from the thermometer, but the last figure is an estimate and its value is therefore uncertain.

The uncertainty associated with an instrument reading is sometimes recorded alongside the measured value. Thus, if x represents a measured value and δ represents the uncertainty in the reading, the final result should be recorded as $x \pm \delta$.

Example 4.2. The uncertainty in the thermometer described in Example 4.1 is 0.05 °C. The temperature reading of 25.30 °C should therefore be written as 25.30 ± 0.05 °C. (This notation gives an estimated bound on the *magnitude* of the error; it says nothing about the sign of the error. In fact, we know that in this case the temperature cannot exceed 25.3 °C.) Similarly, the other reading should be written as 25.20 ± 0.05 °C.

4.2 ERRORS

Measured data always include some element of error. Sometimes the predominant source of error is an improperly adjusted instrument, or the use of the wrong formula or procedure. Such errors are called *consistent* errors.

The possibility of introducing consistent errors into a set of measured data can never be completely eliminated. The likelihood of such errors occurring can be minimized, however, by careful experimental technique. In many cases a consistent error can be detected by measuring a known value as a check point.

Example 4.3. A student has been using a thermocouple to measure the freezing point of a salt solution at various salt concentrations. The following data were obtained.

Concentration, %	5	10	15	20
Freezing Point, °C	−5	−8	−12	−18

The data were then plotted on an arithmetic graph, as shown by the solid curve in Fig. 4-1. When this curve was extrapolated to zero percent salt (pure water), a freezing point of -2 °C was indicated. Since water is known to freeze at 0 °C, the student suspected that the thermocouple was improperly calibrated. He therefore recalibrated the thermocouple, and found that all the temperatures should be 2 °C higher than the values presented above. The dashed curve in Fig. 4-1 indicates the corrected values.

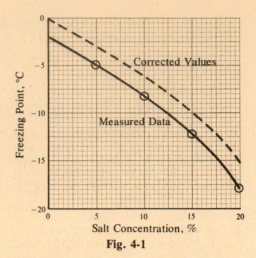

Fig. 4-1

Carefully obtained data will not usually include consistent errors. However, there is another type of error that is present in *all* measured data, at least to some degree. This is the *random* error. Such errors are caused by random fluctuations in the measuring instrument, or by variations in the perception or the interpretations of the observer.

Although random errors can never be completely eliminated, their impact can usually be reduced to a tolerable level through careful experimental technique. The use of simple data reduction procedures, such as the averaging of multiple data points, can be very helpful.

Example 4.4. The electrical resistance of a resistor rated at 10 Ω has been carefully measured six times, the results being 9.83, 10.08, 10.27, 9.91, 10.03, 9.86 Ω. If we calculate the average of the six data points, we obtain

$$R_{\text{avg}} = \frac{9.83 + 10.08 + 10.27 + 9.91 + 10.03 + 9.86}{6} = \frac{59.98}{6} \approx 10.00 \, \Omega$$

Thus the average value is exactly as it should be (to two decimals), despite the variation in the individual measurements.

The above example illustrates the three important characteristics of random errors; namely, that *small errors are more likely to occur than large errors*, that *very large errors are highly unlikely*, and that *positive and negative errors are equally likely and therefore tend to cancel one another*. The success of the above averaging procedure is explained by the last of these characteristics.

Once an average value has been established for a series of measurements, it is helpful to calculate the deviations of the individual values from the average. Each deviation is approximately equal to the random error associated with that measurement. The deviations can be calculated as

$$d_i = x_i - x_{\text{avg}}$$

where d_i = deviation of the *i*th measurement from the average

        x_i = the *i*th measured value

    x_{avg} = the average value

Example 4.5. Calculate the deviations from the average for the data presented in Example 4.4, and identify the largest (in magnitude) deviation.

The average value obtained in Example 4.4 was 10.00 Ω, to two-decimal accuracy. Thus, $x_{\text{avg}} = 10.00 \, \Omega$. The individual deviations, calculated from the above formula, are

$$9.83 - 10.00 = -0.17 \, \Omega$$
$$10.08 - 10.00 = 0.08$$
$$10.27 - 10.00 = 0.27$$
$$9.91 - 10.00 = -0.09$$
$$10.03 - 10.00 = 0.03$$
$$9.86 - 10.00 = -0.14$$

The largest deviation occurs in the third measurement (i.e. $d_3 = 0.27 \, \Omega$), indicating that the third measurement probably contains the greatest error.

In many situations an accurate value for the quantity being measured will be known in advance. This value may be the result of a large number of previous measurements, or it may be the result of a calculation based upon theory. In either case we can interpret such a value as an "accepted true value."

If an accepted true value is available, the *absolute error* in a measured value is defined as

$$e_i = x_i - x$$

where e_i = absolute error in the ith measurement
          x_i = the ith measured value
          x = the accepted true value

If an accepted true value is not known, then x_{avg}, the average of the measured values, is usually used in its place.

Example 4.6. A student has obtained several measurements for the density of steel, as shown below. Determine the absolute error associated with each measured value, if the accepted true value is 7.84×10^3 kg/m^3.

Measurement No.	1	2	3	4	5
Density, 10^3 kg/m^3	7.86	7.84	7.83	7.84	7.87

The average of the measured values is 7.85×10^3 kg/m^3. However, the absolute errors are based upon the accepted true value rather than the average. Thus, the absolute errors, in units of 10^3 kg/m^3, are as follows:

$$7.86 - 7.84 = \quad 0.02$$
$$7.84 - 7.84 = \quad 0$$
$$7.83 - 7.84 = -0.01$$
$$7.84 - 7.84 = \quad 0$$
$$7.87 - 7.84 = \quad 0.03$$

The *relative error* is defined as the ratio of the absolute error to the accepted true value, i.e.

$$r_i = \frac{e_i}{x} = \frac{x_i - x}{x}$$

where r_i is the relative error in the ith measurement. Relative error is often preferred to absolute error because it allows the error to be expressed in relation to the quantity being measured. Multiplied by 100, the relative error becomes the *percent error*.

Example 4.7. For the measured values presented in Example 4.6 (where $x = 7.84 \times 10^3$ kg/m^3) the relative errors and the corresponding percent errors are calculated in Table 4-1. Notice that the relative and percent errors are dimensionless.

Table 4-1

Measurement No.	Density, 10^3 kg/m^3	Relative Error	Percent Error, %
1	7.86	$(7.86 - 7.84)/7.84 = \quad 0.0026$	0.26
2	7.84	$(7.84 - 7.84)/7.84 = \quad 0$	0
3	7.83	$(7.83 - 7.84)/7.84 = -0.0013$	-0.13
4	7.84	$(7.84 - 7.84)/7.84 = \quad 0$	0
5	7.87	$(7.87 - 7.84)/7.84 = \quad 0.0038$	0.38

4.3 THE ARITHMETIC AVERAGE

The computation of a simple arithmetic average has been carried out in Example 4.4. For n data points $x_1, x_2, ..., x_n$, we have

$$x_{\text{avg}} = \frac{x_1 + x_2 + \cdots + x_n}{n}$$

This quantity is a "simple" average because each measured value contributes equally, i.e. each measured value has the same "weight".

In many situations we may wish to place a greater emphasis on certain of the data points when calculating an average. This can be accomplished by means of the formula

$$x_{\text{avg}} = f_1 x_1 + f_2 x_2 + \cdots + f_n x_n$$

where $f_1, f_2, ..., f_n$ are *weighting factors*, which must satisfy the conditions $f_i \geq 0$ for $i = 1, 2, ..., n$, and

$$f_1 + f_2 + \cdots + f_n = 1$$

The greater its weighting factor, the greater the influence of a data point on the average. The simple arithmetic average is obtained in the special case

$$f_1 = f_2 = \cdots = f_n = \frac{1}{n}$$

Example 4.8. A student's scores in four exams are 60, 75, 65, and 90. Determine his overall semester average if the first exam contributes 20 percent to the overall average, the second and third exams each contribute 25 percent, and the last exam contributes 30 percent.

The data points for this problem are $x_1 = 60$, $x_2 = 75$, $x_3 = 65$, and $x_4 = 90$. The corresponding weighting factors (which must be expressed as decimals) are $f_1 = 0.20$, $f_2 = 0.25$, $f_3 = 0.25$, and $f_4 = 0.30$. Thus,

$$x_{\text{avg}} = (0.20)(60) + (0.25)(75) + (0.25)(65) + (0.30)(90) = 74$$

It is interesting to observe that the student's overall average would have been only 72.5 if a simple arithmetic average had been calculated.

4.4 STATISTICAL DATA CHARACTERISTICS

There are several different statistical quantities that can be used to characterize a set of data. Some of these, such as the mean, the median, and the mode, are representative "*central*" points about which the data tend to be clustered. Others, such as the variance and the standard deviation, characterize the degree of *dispersion* (scatter) in the data.

Mean

The *mean value*, $\bar{x}$, of a data set has the property that

$$n\bar{x} = x_1 + x_2 + \cdots + x_n \qquad \text{or} \qquad \bar{x} = \frac{x_1 + x_2 + \cdots + x_n}{n} = x_{\text{avg}}$$

Thus the mean value of a set of data is equal to the simple arithmetic average.

If we have several sets of data, where the ith set contains n_i values and has mean $\bar{x}_i$, an overall mean can be defined as

$$\bar{x} = \frac{n_1 \bar{x}_1 + n_2 \bar{x}_2 + \cdots + n_k \bar{x}_k}{n_1 + n_2 + \cdots + n_k}$$

This is simply a weighted arithmetic average, where each weighting factor represents the fraction of total data values in the corresponding data set.

Example 4.9. Table 4-2 summarizes three sets of tests that were conducted to determine the force that will cause a steel structural component to fail. Determine the overall mean failure point.

Table 4-2

Test Set	Testing Agency	Number of Measurements	Mean Failure Point, N
1	Steel Fabricator	30	4783
2	University	12	4620
3	Independent Testing Laboratory	18	4705

$$\bar{x} = \frac{(30)(4783) + (12)(4620) + (18)(4705)}{30 + 12 + 18} = 4727 \text{ N}$$

In Section 4.5 we will encounter a different type of situation, in which the mean is again equal to a weighted arithmetic average.

Median

If the values in a set of data are arranged in ascending order, the *median* will be the midmost value.

Example 4.10. An automobile engine requires a casting whose size and weight must be carefully controlled. The weights (masses) of nine samples, taken at random from the production line, are 11.33, 11.27, 11.38, 11.30, 11.29, 11.30, 11.34, 11.31, and 11.32 kg. Determine the median of the data.

We begin by rearranging the values into ascending order. Thus, the data set becomes

$$11.27, \ 11.29, \ 11.30, \ 11.30, \ 11.31, \ 11.32, \ 11.33, \ 11.34, \ 11.38$$

The fifth value, 11.31 kg, is the median, since it is surrounded by four values above and below.

The median will always coincide with one of the measured values if the total number of values is odd, as in the above example. If the total number of values is even, however, then the median will fall between two measured values; it is taken as the arithmetic average of these two middle values. Under this convention, every set of data will have a single (unique) mean and a single (unique) median.

Mode

The *mode* of a set of data is the value that occurs with the greatest frequency.

Example 4.11. Determine the mode of the data presented in Example 4.10.

In this data set the value 11.30 appears twice, whereas all other values appear only once. Therefore the mode is 11.30 kg.

Some data sets will have more than one mode.

Example 4.12. Determine the mode of the following data: 5, 2, 4, 12, 10, 12, 5, 8.

In this data set there are *two* values that appear twice, 5 and 12. All other values appear only once. Therefore the values 5 and 12 each represent a mode.

A data set that has only one mode, as in Example 4.11, is said to be *unimodal*. The data set presented in Example 4.12 is *bimodal*, since it contains two different values for the mode. In general, a set of data that contains more than one mode is said to be *multimodal*.

The mean, the median, and the mode all supply useful information about a set of data.

Example 4.13. A freshman engineering class contains 20 students. At the end of the term the students had earned the following final grades:

15	30	55	70	80
20	30	60	75	80
25	30	65	75	85
25	40	70	75	90

The class average (mean) is 54.8, although half the students had grades above 62.5 (the median). The grade distribution is bimodal, with one group of grades clustered around 30 and another group around 75.

Variance

The *variance* of a data set indicates the amount of dispersion of the individual data values about the mean. For the set $x_1, x_2, ..., x_n$, the variance is defined as

$$v = \frac{(x_1 - \bar{x})^2 + (x_2 - \bar{x})^2 + \cdots + (x_n - \bar{x})^2}{n}$$

where $\bar{x}$ is the mean of the set. Since each term in the numerator is nonnegative and $n > 0$, we see that the variance is always nonnegative, i.e. $v \geq 0$. Moreover, v will equal zero only if all the data values are the same, i.e.

$$x_1 = x_2 = \cdots = x_n = \bar{x}$$

The greater the variation about the mean, the larger will be the value for v.

The above formula can be rearranged to give

$$v = \frac{x_1^2 + x_2^2 + \cdots + x_n^2}{n} - \bar{x}^2$$

(Problem 4.11). This formula is preferred for computational purposes, since it is easier to work with than the earlier formula. In either case the value for $\bar{x}$ must be expressed as accurately as possible.

Example 4.14. Calculate the variance for the data in Example 4.10.

The mean is

$$\bar{x} = \frac{11.33 + 11.27 + 11.38 + 11.30 + 11.29 + 11.30 + 11.34 + 11.31 + 11.32}{9} = 11.315\,556 \text{ kg}$$

and so

$$v = \frac{(11.33)^2 + (11.27)^2 + (11.38)^2 + (11.30)^2 + (11.29)^2 + (11.30)^2 + (11.34)^2 + (11.31)^2 + (11.32)^2}{9} - (11.315\,556)^2$$

$$= 0.000\,903\,5 \text{ kg}^2$$

Thus the variance has a value of about 10^{-3} kg². (Note that a more accurate value, $v = 0.000\,913\,6$ kg², is obtained from the first formula.) The calculated value of the variance would have been considerably in error if we had used the rounded value, $\bar{x} = 11.32$ kg (see Problem 4.12).

Standard Deviation

The *standard deviation*, s, is defined as the square root of the variance, i.e.

$$s = \sqrt{v} = \sqrt{\frac{x_1^2 + x_2^2 + \cdots + x_n^2}{n} - \bar{x}^2}$$

It provides essentially the same information as the variance; that is, it indicates the degree of dispersion about the mean. However, its magnitude more closely approximates the magnitudes of the individual deviations. Further, it has the same units as the individual data points.

Example 4.15. Calculate the standard deviation for the data presented in Example 4.10.
From Example 4.14 the variance is $v = 0.000\,903\,5$ kg². Therefore

$$s = \sqrt{9.035 \times 10^{-4}} = 3.006 \times 10^{-2} \text{ kg} \approx 0.03 \text{ kg}$$

This value is representative of the individual deviations about the mean, which range from 0 to 0.06 kg in magnitude.

Some authors divide by $n - 1$ rather than n when calculating the variance and the standard deviation. This is particularly common in statistics texts. We will not pursue this point further in this book. It may be mentioned in passing, however, that if n is modestly large, there is very little difference between the values obtained using n and $n - 1$.

4.5 GROUPED DATA

It is often desirable to present a set of data in terms of the numbers of events occurring in several adjacent intervals. These numbers specify the *distribution* of the data.

Example 4.16. During the month of January, the average daily temperatures in Fairweather, Florida, were as shown in Table 4-3.

Table 4-3

Day of Month	Temperature, °C	Day of Month	Temperature, °C	Day of Month	Temperature, °C
1	17	11	18	21	23
2	15	12	20	22	24
3	11	13	22	23	21
4	13	14	23	24	18
5	8	15	21	25	19
6	9	16	26	26	20
7	6	17	30	27	17
8	10	18	28	28	15
9	12	19	27	29	17
10	16	20	30	30	14
				31	12

Find the distribution of the data over intervals of 5 °C.
The proper grouping of the data is shown below.

Interval	Number of Days
6–10 °C	4
11–15	7
16–20	9
21–25	6
26–30	5
TOTAL	31

When the data are arranged in this manner, it is clear that most of the average daily temperatures fall between 11 and 25 °C. Moreover, temperatures between 16 and 20 °C are prevalent.

In many situations it may be preferable to present grouped data in terms of *relative frequency* rather than number of data points (see Example 3.7). We have

$$f_i = \frac{n_i}{n}$$

where f_i = the relative frequency for the ith interval, $i = 1, 2, \ldots, k$

n_i = the number of data points in the ith interval

n = the total number of data points

It is easy to see that $f_i \geq 0$ and that $f_1 + f_2 + \cdots + f_k = 1$. Thus, the relative frequencies can be interpreted as *weighting factors* (for the intervals, not for the individual data points as in Section 4.3).

Example 4.17. The relative frequencies for the grouped data obtained in Example 4.16 are calculated in Table 4-4.

Table 4-4

Interval	Relative Frequency (Fraction)	Relative Frequency (Percentage)
6–10 °C	4/31 = 0.129	12.9%
11–15	7/31 = 0.226	22.6
16–20	9/31 = 0.290	29.0
21–25	6/31 = 0.194	19.4
26–30	5/31 = 0.161	16.1
TOTALS	1.000	100.0%

Thus about 13% of the average daily temperatures fall into the interval 6–10 °C, 23% fall into the interval 11–15 °C, and so on.   The interval with the greatest frequency is clearly 16–20 °C.

4.6 FREQUENCY DISTRIBUTIONS

A set of grouped data is often referred to as a *frequency distribution*, since it indicates the frequency with which the data fall into each constituent interval.   Frequency distributions can be represented by histograms (see Section 3.2) and, in some cases, analytically.   Important characteristics, such as symmetry, location of modes, and degree of dispersion, can readily be seen from the histogram.

Example 4.18. Another and more extensive set of data describing the average daily temperature in Fairweather, Florida, during the month of January has been obtained from the U.S. Weather Bureau.   The frequency distribution, in terms of 2 °C temperature intervals, is shown below.

Interval, °C	6–8	8–10	10–12	12–14	14–16	16–18	18–20	20–22
Relative Frequency, %	0.6	3.8	9.0	12.2	13.5	12.8	11.7	10.2
Interval	22–24	24–26	26–28	28–30	30–32	32–34	34–36	
Relative Frequency	8.3	6.4	4.5	3.2	1.9	1.3	0.6	

A histogram of the data is shown in Fig. 4-2.   From this figure it is clear that the distribution is unimodal and asymmetric about the mode, favoring the higher temperatures (this type of asymmetry is called *positive skewness*).   The mode itself appears to be located somewhere between 14 and 16 °C.   The data are dispersed over a 30 °C range, running from 6 °C to 36 °C.

A frequency distribution will often represent some process consisting of a number of randomly occurring events.   If the data set is large enough to be truly representative of the process, then each of the frequencies can be interpreted as the *probability* that a random event will fall into the corresponding interval.

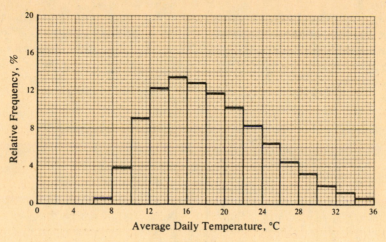

Fig. 4-2

Example 4.19.   What is the probability that, for a day in January in Fairweather, Florida, the average temperature will be between 16 and 18 °C?   (See the relative frequencies given in Example 4.18.)

Supposing that the data set used to compute the relative frequencies is large enough to be truly representative of Fairweather's temperature pattern during the month of January, there is a 12.8 % probability that the average daily temperature will be between 16 and 18 °C, since this interval has a relative frequency of 12.8 % (i.e. 12.8 % of the recorded temperatures fall into this interval).

The mean value of a frequency distribution is equal to the weighted arithmetic average, where the (fractional) relative frequencies serve as weighting factors, and the midpoints of the corresponding intervals are the individual data values.   Mathematically, this can be expressed as

$$\overline{x} = f_1 x_1 + f_2 x_2 + \cdots + f_k x_k$$

Example 4.20.   Determine the mean value for the frequency distribution presented in Example 4.18.

The desired value is easily obtained using the above formula.   Thus,

$$\overline{x} = (0.006)(7) + (0.038)(9) + (0.09)(11) + (0.122)(13) + \cdots + (0.013)(33) + (0.006)(35)$$
$$= 18.406 \approx 18.4 \text{ °C}$$

A more accurate value would have been obtained if we could have calculated the simple arithmetic average of the raw (ungrouped) data.   But, as is often the case, the raw data were not available.

The median of a frequency distribution can best be found from the corresponding cumulative distribution.   We will see how this value is obtained in Section 4.7.

The variance can be determined from the expression

$$v = (f_1 x_1^2 + f_2 x_2^2 + \cdots + f_k x_k^2) - \overline{x}^2$$

where f_i represents the (fractional) relative frequency of the ith interval, x_i represents the midpoint of that interval, and x represents the mean value of the frequency distribution.   The standard deviation is defined as the square root of the variance, $s = \sqrt{v}$,   as before.

Example 4.21.   Determine the variance and the standard deviation for the frequency distribution presented in Example 4.18.

The variance can be obtained from the above formula using the value of $\overline{x}$ determined in Example 4.20, i.e. $\overline{x} = 18.406$ °C.   Thus,

$$v = [(0.006)(7)^2 + (0.038)(9)^2 + (0.019)(11)^2 + (0.122)(13)^2 + \cdots + (0.013)(33)^2 + (0.006)(35)^2] - (18.406)^2$$
$$= 34.075\,164 \approx 34.1 \text{ (°C)}^2$$

The standard deviation can now be obtained as

$$s = \sqrt{34.075\,164} = 5.837\,394 \approx 5.84 \text{ °C}$$

In many situations a frequency distribution obtained from an extensive data base will be expressed in terms of a large number of narrow intervals. If the resulting histogram is "smooth" (i.e. does not contain "peaks" and "valleys"), as in Fig. 4-3, then it is very natural to pass a smooth, continuous curve through the midpoints of the intervals. Figure 4-3 shows such a curve. The accuracy with which the curve represents the frequency distribution will increase as the total number of data values becomes larger and the width of the intervals becomes smaller.

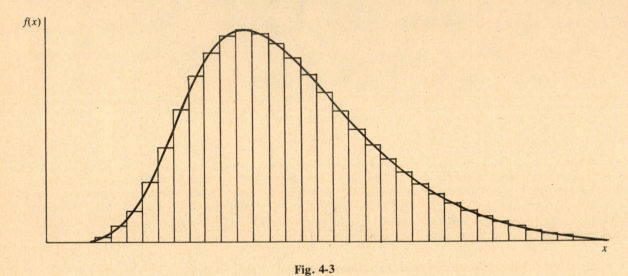

Fig. 4-3

Usually, when a frequency distribution is represented by a continuous curve, the ordinate values are adjusted so that the total area under the curve has a value of unity, i.e. the scale along the ordinate is adjusted such that

$$\int_{-\infty}^{\infty} f(x)\, dx = 1$$

The frequency function $f(x)$, which must also obey the condition $f(x) \geq 0$, is then called a *probability density function*. If the frequency distribution describes the occurrence of events in a particular sort of random process, the function $f(x)$ is known as a *theoretical probability density function*. One such density function will be examined in Section 4.8.

4.7 CUMULATIVE DISTRIBUTIONS

It is often more useful to present a set of grouped data in terms of a cumulative distribution rather than a frequency distribution. The individual terms are easily obtained as the partial sums of the relative frequencies. (See Examples 3.7 and 3.8.)

If the grouped data represent outcomes of a certain random process (e.g. selecting at random a day of the month and measuring the temperature on that day), then the ordinate of the cumulative distribution corresponding to the ith interval can be interpreted as the probability that an outcome will not exceed x_i, where x_i is the right boundary of the ith interval. This interpretation will, of course, be valid only if the data set is large enough to be truly representative of the process.

Example 4.22. What is the probability that, for a day in January in Fairweather, Florida, the average temperature will be less than or equal to 18 °C?

Figure 4-4 shows the cumulative distribution of the temperature data provided in Example 4.18. From this figure we see that the ordinate value corresponding to 18 °C (that is, to the right boundary of the interval 16–18 °C) is about 52% (actually, 51.9%). Thus, there is a 52% probability that on a day in January the average temperature in Fairweather, Florida, will be less than or equal to 18 °C.

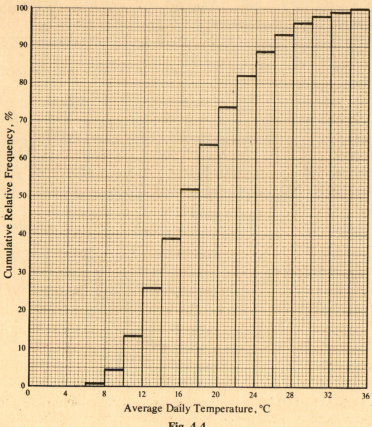

Fig. 4-4

We have already seen that the mean, the mode(s), the variance and the standard deviation for a set of data can easily be determined from the frequency distribution.   The median, however, is best obtained from the cumulative distribution, using the linear interpolation formula

$$\tilde{x} = x_{m-1} + \frac{50 - y_{m-1}}{y_m - y_{m-1}}(x_m - x_{m-1})$$

where $\tilde{x}$ = the median value
x_{m-1} = the left boundary of the mth interval
x_m = the right boundary of the mth interval
y_{m-1} = the ordinate value for the $(m-1)$st interval
y_m = the ordinate value for the mth interval

In the above formula the mth interval is the interval having the smallest ordinate value which is greater than or equal to 50% (i.e. the *leftmost* interval whose ordinate value is greater than or equal to 50%).

Example 4.23.   Calculate the median of the temperature data in Example 4.18.
The interval 16–18 °C is the leftmost interval whose ordinate value (51.9 %) is greater than or equal to 50% (see Fig. 4-4).   Thus, using the above formula,

$$\tilde{x} = 16 + \frac{50 - 39.1}{51.9 - 39.1}(18 - 16) = 17.7 \text{ °C}$$

In Section 4.6 we saw how the histogram of a frequency distribution passed over into a smooth curve, the probability density function $f(x)$, as the intervals increased in number and decreased in width.   Similarly, the graph of the corresponding cumulative distribution passes over into the *cu-*

mulative distribution function $F(x)$.   (The only difference is that the approximating curve is passed through the *right boundary* of each interval in the graph of the cumulative distribution.   See Fig. 4-5.)   The function $F(x)$ is nondecreasing in x, and has the limiting values

$$F(-\infty) = 0 \qquad F(+\infty) = 1 \text{ or } 100\% \quad \text{(depending on the ordinate scale)}$$

The relation between $F(x)$ and $f(x)$ (when fractional ordinates are used) is

$$F(x) = \int_{-\infty}^{x} f(x') \, dx'$$

that is, the distribution function is the integral of the density function.   (Recall that an integral actually represents the limiting value of a cumulative sum.)

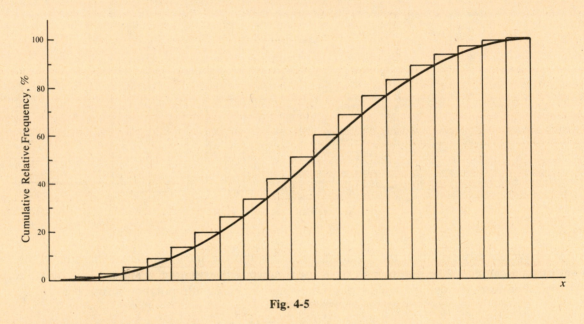

Fig. 4-5

4.8   THE NORMAL DISTRIBUTION

Many different theoretical distribution functions have been developed to describe certain types of processes involving random events.   Of these, one is of particular importance in science and engineering.   This is the *normal*, or *Gaussian*, distribution, which can be used to characterize the variation in measured data for a great many different physical situations.   For example, the heights and weights of people, the variations in manufacturing tolerances, and, in some situations, the occurrence of random errors all tend to be normally distributed.

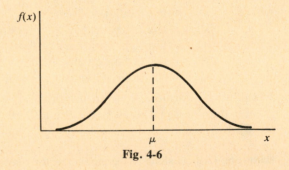

Fig. 4-6

The *normal curve* (the graph of the probability density function) is bell-shaped and symmetric, as illustrated in Fig. 4-6.   Although there are some other theoretical density functions that have a similar appearance, the normal density function is usually used to represent data whose frequency distribution is approximately symmetric and bell-shaped.

Example 4.24.   A factory manufactures integrated circuits ("chips") for electronic calculators.   Table 4-5 indicates the number of defective chips manufactured during each of several 8-hour shifts.

Table 4-5

Defects per Shift	Number of Occurrences	Relative Frequency, %
0–2	0	
3–4	2	3.6
5–6	4	7.3
7–8	8	14.5
9–10	12	21.8
11–12	13	23.6
13–14	9	16.4
15–16	5	9.1
17–18	2	3.6
TOTALS	55	99.9%

Plot the relative frequency for each interval against the corresponding interval midpoint. What can be inferred about the nature of the distribution?

Figure 4-7 shows a plot of the data. There are actually too few data points, and too much scatter, to determine a precise, smooth curve. It is plausible that the frequencies can be represented by a symmetric bell-shaped curve, as indicated by the dashed curve in Fig. 4-7. From this plot it is not clear, however, that the data are necessarily governed by a normal distribution.

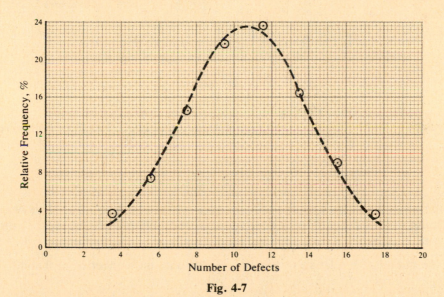

Fig. 4-7

In Example 3.18 we constructed the cumulative distribution for this same set of data and plotted the cumulative distribution on probability paper (Fig. 3-25). The resulting straight line led us to conclude that the frequency data can indeed be represented by a normal density function. Thus, we see that plotting a cumulative distribution on probability paper is a much more conclusive test than plotting relative frequencies. We will say more about this later in this section.

The normal distribution has the following important characteristics.

1. The density function represented by the normal curve is

$$f(x) = \frac{1}{\sigma \sqrt{2\pi}} e^{-[(x-\mu)/\sigma]^2/2}$$

where μ is the mean and σ is the standard deviation.   (Note that these two parameters apply to the normal *curve*—not the *data* represented by the curve.   Ordinarily, μ and σ will differ somewhat from the sample mean, $\bar{x}$, and the sample standard deviation, s, obtained directly from the data.)

2. The normal curve is symmetric about the mean.

3. The median and the mode have the same value as the mean.

4. The area under the normal curve is equal to unity, i.e.

$$\int_{-\infty}^{\infty} f(x)\, dx = 1$$

(This is true of all theoretical density functions.)

5. If a set of random events is governed by a normal distribution with mean μ and standard deviation σ, then 68.2% of the events will have corresponding values of x between $\mu - \sigma$ and $\mu + \sigma$.   Furthermore, 95.4% of the events will have corresponding values of x between $\mu - 2\sigma$ and $\mu + 2\sigma$, and 99.7% will have corresponding values of x between $\mu - 3\sigma$ and $\mu + 3\sigma$.   These percentages correspond to the portions of the area under the curve shown in Fig. 4-8.

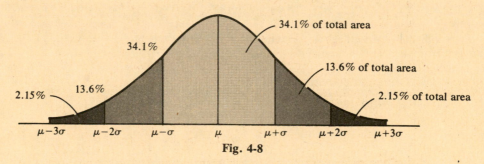

34.1% of total area

34.1%

13.6% of total area

2.15%   13.6%

2.15% of total area

$\mu-3\sigma$ $\mu-2\sigma$ $\mu-\sigma$ μ $\mu+\sigma$ $\mu+2\sigma$ $\mu+3\sigma$

Fig. 4-8

Properties 2, 3, and 5 are especially useful in data analysis, as discussed below.

We have already learned that a cumulative normal distribution, i.e. the cumulative distribution function

$$F(x) = \int_{-\infty}^{x} f(x')\, dx' = \frac{1}{\sigma\sqrt{2\pi}} \int_{-\infty}^{x} e^{-[(x' - \mu)/\sigma]^2/2}\, dx'$$

can be represented by a straight line on a probability graph (see Section 3.5).   This offers us a simple method for fitting a normal distribution to a set of data (provided, of course, that the data can be represented with reasonable accuracy by such a distribution).   The procedure is as follows.

1. Plot the cumulative relative frequencies of the data on a sheet of probability graph paper, and pass a straight line through the data points.

2. Read the abscissa corresponding to an ordinate of 50% (or 0.5, if the ordinates range from 0 to 1).   This will be the mean, μ (as well as the median and the mode).

3. Read the abscissa corresponding to an ordinate of $50 - 34.1 = 15.9\%$.   This value equals $\mu - \sigma$.   Since μ has already been determined, we can now obtain $\sigma = \mu - (\mu - \sigma)$.

With μ and σ known, the normal density function,

$$f(x) = \frac{1}{\sigma\sqrt{2\pi}} e^{-[(x - \mu)/\sigma]^2/2}$$

becomes an explicit, known function of x.

Example 4.25.   Fit a normal curve to the data of Example 4.24.

Figure 4-9 reproduces Fig. 3-25, where the cumulative relative frequencies of the data were originally plotted.

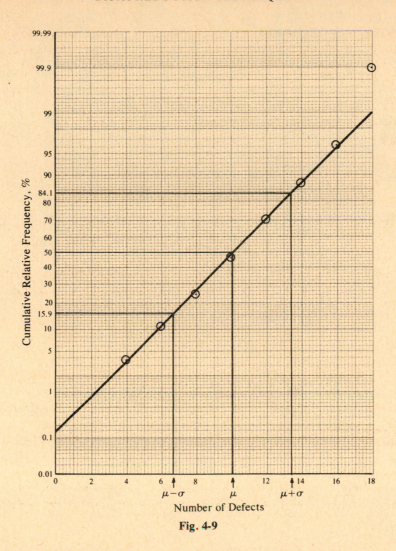

Fig. 4-9

From Fig. 4-9 we see that the abscissa corresponding to an ordinate of 50% is 10.1. Therefore $\mu =$ 10.1 (hence the mean, the median, and the mode all have values of 10.1). Furthermore, the abscissa corresponding to an ordinate of 15.9% is 6.7. Therefore

$$\sigma = 10.1 - 6.7 = 3.4$$

We can also obtain the standard deviation by reading the abscissa corresponding to an ordinate of 84.1%. This yields $\mu + \sigma = 13.5$, whence

$$\sigma = 13.5 - 10.1 = 3.4$$

as before.

The desired normal density function can now be written as

$$f(x) = \frac{1}{3.4\sqrt{2\pi}} e^{-[(x-10.1)/3.4]^2/2} = 0.117\,336\,e^{-[(x-10.1)/3.4]^2/2}$$

It should be understood that a normal curve which has been obtained as in Example 4.25 may have entirely different ordinate values than the original frequency distribution. The reason for this apparent inconsistency is that the area under the normal curve must equal unity, a condition not necessarily satisfied by the original frequency data. This discrepancy can be removed by multiplying all the original frequencies by an appropriate scale factor (see Problem 4.18).

4.9 CURVE-FITTING PROCEDURES

We have already seen that certain mathematical equations can be fitted to measured data by plotting the data on an appropriate type of graph paper (e.g. semilog, log-log, probability) and then passing a straight line through the data points.   Of course, the method will be valid only if the data points define a reasonably straight line.   But even then the method is subjective; no two people will draw exactly the same straight line through the data.

Another approach to the problem of fitting a curve to a set of data points is to use the *method of least squares*.   This method is consistent, in the sense that it provides a systematic procedure for fitting a unique curve to a set of data.   There is no personal judgment involved.   Moreover, the procedure can easily be implemented with a computer or an electronic calculator.

Suppose that we wish to fit a curve $y = f(x)$ to a set of M data points (x_1, y_1), (x_2, y_2), ..., (x_M, y_M).   The method of least squares is based upon the concept of minimizing the sum of the squared errors, i.e.

Minimize $S = e_1^2 + e_2^2 + \cdots + e_M^2$

where e_i, the ith error, is the difference between the observed value y_i and the ordinate of the fitting curve at x_i, as illustrated in Fig. 4-10.

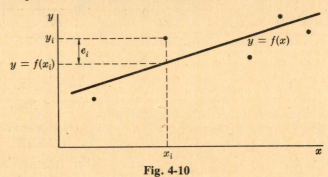

Fig. 4-10

The method is commonly applied to straight lines, polynomials, power functions, and exponential functions.   In each case the method requires solving a set of simultaneous, linear algebraic equations, where the unknown quantities are the *constants* in the equation for the curve.   (An efficient method for solving simultaneous, linear algebraic equations will be presented in Chapter 5.)

Least-squares Straight Line

Suppose that we wish to pass a straight line, $y = c_1 + c_2 x$,   through a set of M data points.   The method of least squares leads to the following two equations for the unknown constants c_1 and c_2 (see Problem 4.19):

$$Mc_1 + \left(\sum_{i=1}^{M} x_i \right) c_2 = \sum_{i=1}^{M} y_i$$

$$\left(\sum_{i=1}^{M} x_i \right) c_1 + \left(\sum_{i=1}^{M} x_i^2 \right) c_2 = \sum_{i=1}^{M} x_i y_i$$

where Σ indicates summation, as

$$\sum_{i=1}^{M} x_i = x_1 + x_2 + \cdots + x_M$$

To prevent large numerical errors it is essential that the summation terms be evaluated as accurately as possible.

Example 4.26. Several measured data points indicating the shear force in a beam (F) as a function of distance from the left end (x) are shown below.   Pass a straight line through the data points, using the method of least squares.

x, cm	10	20	30	40	50	60	70	80
F, N	4	13	20	22	27	38	42	45

The first step in carrying out the calculations is to form Table 4-6.

Table 4-6

x	x^2	F	xF
10	100	4	40
20	400	13	260
30	900	20	600
40	1 600	22	880
50	2 500	27	1 350
60	3 600	38	2 280
70	4 900	42	2 940
80	6 400	45	3 600
TOTALS 360	20 400	211	11 950

Substituting the cumulative sums into the least-squares equations for a straight line, we obtain

$$8c_1 + 360c_2 = 211$$
$$360c_1 + 20\,400c_2 = 11\,950$$

These two equations can be solved by elementary methods (e.g. substitution) or by the method presented in Chapter 5. In any event, the solution is $c_1 = 0.071\,429$, $c_2 = 0.584\,524$. Therefore the desired linear fit is

$$F = 0.0714 + 0.585x$$

Figure 4-11 shows a plot of this equation, along with the data points.

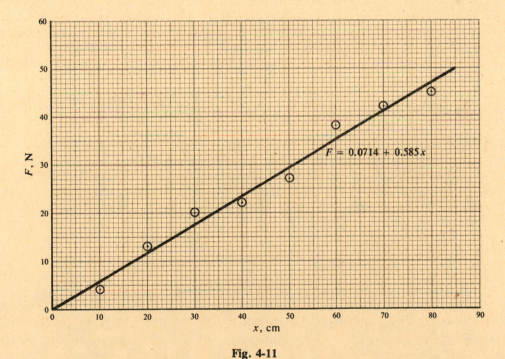

Fig. 4-11

The method of least squares is particularly useful in problems such as this, where the data points show an appreciable amount of scatter. The use of subjective curve-fitting procedures (e.g. "eyeballing" a curve through the data) can thereby be avoided.

Least-squares Polynomial

Now let us consider a more general case, where we fit the polynomial

$$y = c_1 + c_2x + c_3x^2 + \cdots + c_{n+1}x^n$$

to a set of M data points. (The straight-line fit corresponds to $n = 1$.) We can determine the coefficients $c_1, c_2, \ldots, c_{n+1}$ by solving the system of $n+1$ simultaneous equations:

$$Mc_1 + \left(\sum_{i=1}^{M} x_i\right)c_2 + \left(\sum_{i=1}^{M} x_i^2\right)c_3 + \cdots + \left(\sum_{i=1}^{M} x_i^n\right)c_{n+1} = \sum_{i=1}^{M} y_i$$

$$\left(\sum_{i=1}^{M} x_i\right)c_1 + \left(\sum_{i=1}^{M} x_i^2\right)c_2 + \left(\sum_{i=1}^{M} x_i^3\right)c_3 + \cdots + \left(\sum_{i=1}^{M} x_i^{n+1}\right)c_{n+1} = \sum_{i=1}^{M} x_i y_i$$

$$\cdots\cdots\cdots\cdots\cdots\cdots\cdots\cdots\cdots\cdots\cdots\cdots\cdots\cdots\cdots$$

$$\left(\sum_{i=1}^{M} x_i^n\right)c_1 + \left(\sum_{i=1}^{M} x_i^{n+1}\right)c_2 + \left(\sum_{i=1}^{M} x_i^{n+2}\right)c_3 + \cdots + \left(\sum_{i=1}^{M} x_i^{2n}\right)c_{n+1} = \sum_{i=1}^{M} x_i^n y_i$$

Problem 4.22 gives a numerical example for $n = 2$ (a parabola).

Least-squares Power Function

Suppose that we wish to pass the power function $y = ax^b$ through a set of M data points. Transforming the equation to

$$Y = c_1 + c_2 X$$

where $Y = \log y$, $X = \log x$, $c_1 = \log a$, $c_2 = b$, we see that the problem reduces to finding the least-squares straight line through the data points $(X_1, Y_1), \ldots, (X_M, Y_M)$. See Problem 4.20 for a numerical example.

Least-squares Exponential Function

Similarly, to pass the exponential function $y = ae^{bx}$ through M data points, we rewrite the equation as

$$Y = c_1 + c_2 X$$

where $Y = \ln y$, $X = x$, $c_1 = \ln a$, $c_2 = b$. Again the problem reduces to finding the least-squares straight line through $(X_1, Y_1), \ldots, (X_M, Y_M)$. See Problem 4.21.

The most suitable type of curve to fit a set of data is usually not known in advance. About all that can be done is to fit several different curves to the data, and then select the curve that best represents the data. This choice can be made by inspecting graphs of the curves and the data points or by computer calculations. In the latter case the computer would actually calculate the sum of the squared errors for each curve, and determine which curve yields the smallest value. We will say more about computer calculations in Chapter 7.

4.10   CORRELATION COEFFICIENT

The *correlation coefficient*, r, for a set of data provides an indication of how well the data can be represented by a straight line. It is a calculated quantity whose value will always lie within the interval $-1 \le r \le 1$. Thus, the data will fall approximately along a straight line if $|r| \approx 1$. (The slope of the line will be positive if $r \approx +1$, and negative if $r \approx -1$.) A straight-line representation will not be valid, however, if the value of r is close to zero.

The correlation coefficient is determined from the formula

$$r = \frac{M\sum_{i=1}^{M} x_i y_i - \left(\sum_{i=1}^{M} x_i\right)\left(\sum_{i=1}^{M} y_i\right)}{\left\{\left[M\sum_{i=1}^{M} x_i^2 - \left(\sum_{i=1}^{M} x_i\right)^2\right]\left[M\sum_{i=1}^{M} y_i^2 - \left(\sum_{i=1}^{M} y_i\right)^2\right]\right\}^{1/2}}$$

where M represents the number of data points, and x_i and y_i represent the observed values of the two variables. Notice that r remains the same when the x_i and the y_i are interchanged; hence it makes no difference which variable we consider to be the independent variable.

Example 4.27. Evaluate the correlation coefficient for the data presented in Example 4.26.
    Here, $M = 8$. The column totals in Table 4-6 give

$$\sum_{i=1}^{8} x_i = 360 \qquad \sum_{i=1}^{8} x_i^2 = 20\,400 \qquad \sum_{i=1}^{8} F_i = 211 \qquad \sum_{i=1}^{8} x_i F_i = 11\,950$$

Moreover,

$$\sum_{i=1}^{8} F_i^2 = 7031$$

Thus, the correlation coefficient is evaluated as

$$r = \frac{8(11\,950) - (360)(211)}{\{[8(20\,400) - (360)^2][8(7031) - (211)^2]\}^{1/2}} = \frac{95\,600 - 75\,960}{[(33\,600)(11\,727)]^{1/2}} = 0.989\,415$$

Since this value is very close to $+1$, we are assured that the data will fall approximately along a straight line whose slope is positive. This conclusion is verified by Fig. 4-11.

Solved Problems

4.1. A voltmeter has a scale that ranges from 0 to 10 V. The scale is divided into intervals of 0.5 V. (*a*) If the position of the needle can be determined to the nearest half interval, what will be the uncertainty associated with the voltmeter when reading this scale? (*b*) Suppose the needle is located near the 8 V division. What voltage should be recorded if the uncertainty in the measurement is to be specified along with the measured value? (*c*) What voltage should be recorded if the needle is located close to the 1.5 V division? (*d*) Now suppose that the needle is located approximately midway between the 5.5 and the 6 V divisions. What voltage should be recorded?

   (*a*) ± 0.25 V   (*b*) 8.0 ± 0.25 V   (*c*) 1.5 ± 0.25 V   (*d*) 5.75 ± 0.25 V. Notice that there is uncertainty in the last *two* digits of this reading. Thus it might be desirable to round up (because of the odd-add rule), resulting in a recorded voltage of 5.8 ± 0.25 V.

4.2. The line voltage supplied by an electrical utility was measured at five different times during the day. The average value was found to be 112.2 V, although the individual deviations were 0.1, 0.7, -1.0, 0.8, and -0.6 V. What were the actual voltage readings?

Measurement No.	Reading
1	$112.2 + 0.1 = 112.3$ V
2	$112.2 + 0.7 = 112.9$
3	$112.2 - 1.0 = 111.2$
4	$112.2 + 0.8 = 113.0$
5	$112.2 - 0.6 = 111.6$

4.3.   Several experiments have been performed to determine the time required for a certain chemical reaction to take place.   The measured values are given below.

Trial No.	1	2	3	4	5	6	7
Time, s	130.4	127.8	123.2	129.5	129.9	127.3	130.8

(*a*)   Calculate the average value of the given measurements.   (*b*)   Calculate the deviations about the average.   (*c*)   Determine which, if any, of the measured values are inaccurate, based upon the calculated deviations.   (*d*)   Reject any inaccurate measurements, then recalculate the average for the remaining data and calculate the individual deviations. (*e*)   Have the calculated average and the individual deviations changed significantly?

(*a*)

$$t_{avg} = \frac{130.4 + 127.8 + 123.2 + 129.5 + 129.9 + 127.3 + 130.8}{7} = 128.4 \text{ s}$$

(*b*)

Trial No.	Deviation
1	$130.4 - 128.4 = 2.0$ s
2	$127.8 - 128.4 = -0.6$
3	$123.2 - 128.4 = -5.2$
4	$129.5 - 128.4 = 1.1$
5	$129.9 - 128.4 = 1.5$
6	$127.3 - 128.4 = -1.1$
7	$130.8 - 128.4 = 2.4$

(*c*)   The deviation associated with the third measurement is more than twice as large (in magnitude) as any other.   It would appear reasonable to reject the third measurement as being inaccurate.

(*d*)

$$t_{avg} = \frac{130.4 + 127.8 + 129.5 + 129.9 + 127.3 + 130.8}{6} = 129.3 \text{ s}$$

Trial No.	Deviation
1	$130.4 - 129.3 = 1.1$ s
2	$127.8 - 129.3 = -1.5$
4	$129.5 - 129.3 = 0.2$
5	$129.9 - 129.3 = 0.6$
6	$127.3 - 129.3 = -2.0$
7	$130.8 - 129.3 = 1.5$

(*e*)   Both the average value and the individual deviations have changed.   The recalculated values are presumably more accurate.

4.4.   Show that the sum of the deviations about the average equals zero (i.e. the deviations cancel one another).

$$d_1 + d_2 + \cdots + d_n = (x_1 - x_{avg}) + (x_2 - x_{avg}) + \cdots + (x_n - x_{avg})$$

$$= (x_1 + x_2 + \cdots + x_n) - nx_{avg}$$

$$= n\left(\frac{x_1 + x_2 + \cdots + x_n}{n}\right) - nx_{avg}$$

$$= nx_{avg} - nx_{avg} = 0$$

In practice, the deviations may not exactly sum to zero because of roundoff errors.

4.5.   A plot of land was determined to be approximately 80 m long.   A more accurate measurement, carried out by a surveyor, gave 81.47 m.   Determine the absolute error, the relative error, and the percent error associated with the approximate measurement, assuming that the surveyor's measurement is, for all practical purposes, error-free.

$$\text{absolute error} = 80 - 81.47 = -1.47 \text{ m}$$

$$\text{relative error} = \frac{80 - 81.47}{81.47} = -0.018\,04$$

$$\text{percent error} = 100 \left(\frac{80 - 81.47}{81.47} \right) = -1.804\%$$

Notice that the absolute error is given to two decimal places, to conform to the true value (the surveyor's measurement).   Similarly, the relative error and the percent error are rounded to four significant figures, in order to be consistent with the true value.

4.6.   A spherical gas storage tank has a measured diameter of 27.4 m.   If the measuring device is accurate to within $\pm 0.5\%$, what will be the volume of the tank?

The actual diameter will fall somewhere within the limits

$$(1 - 0.005)(27.4) = 27.263 \text{ m} \qquad \text{and} \qquad (1 + 0.005)(27.4) = 27.537 \text{ m}$$

The corresponding limits for the tank's volume are

$$V_{\min} = \frac{\pi (27.263)^3}{6} = 10\,610 \text{ m}^3 \qquad \text{and} \qquad V_{\max} = \frac{\pi (27.537)^3}{6} = 10\,933 \text{ m}^3$$

Therefore the volume will be somewhere between 10 600 and 10 900 m³.

4.7.   Suppose that the measured diameter in Problem 4.6 has been found to be 0.4% too low.   What will be the volume of the tank?

The percent error is now based upon the *true* diameter rather than the *measured* diameter.   Therefore the actual tank diameter is calculated as

$$\frac{27.4}{1 - 0.004} = 27.510 \text{ m}$$

Hence the tank volume will be

$$V = \frac{\pi (27.510)^3}{6} = 10\,901 \text{ m}^3$$

4.8.   The arithmetic average of five measurements has been determined to be 47.65.   A sixth measurement, whose value is 55.17, has recently been obtained.   Determine the arithmetic average of all six measured values.

$$x_{\text{avg}} = \frac{5(47.65) + 55.17}{6} = \frac{5}{6}(47.65) + \frac{1}{6}(55.17) = 48.90$$

4.9.   At the end of each term a university determines a "quality point average" (QPA) for each student.   The QPA is actually a weighted average, where the quantities being averaged are the numerical equivalents of the letter grades earned during the term.   The weighting factors are the number of credit hours for each course divided by the total number of credit hours for the term.   Determine the QPA for the freshman engineering student whose grade record is shown below.

Course	Credit Hours	Grade
Calculus I	4	A
English Composition	3	C
Chemistry	4	B
Intro. to Engineering	2	A
Engineering Analysis	2	B
Physical Education	1	D
TOTAL	16	

At this university, an A has a numerical equivalent of 4 points, a B is equivalent to 3 points, etc.

$$QPA = \frac{4(4) + 3(2) + 4(3) + 2(4) + 2(3) + 1(1)}{16}$$

$$= \frac{4}{16}(4) + \frac{3}{16}(2) + \frac{4}{16}(3) + \frac{2}{16}(4) + \frac{2}{16}(3) + \frac{1}{16}(1) = 3.063$$

4.10. Calculate the mean, the median, the mode, the variance, and the standard deviation for the following set of data: $4, -3, 5, 8, 11, -2, 0, -5$.

$$\bar{x} = \frac{4 - 3 + 5 + 8 + 11 - 2 + 0 - 5}{8} = 2.250 \qquad \tilde{x} = \frac{0 + 4}{2} = 2.0$$

Because each data value occurs only once, the mode does not exist.

$$v = \frac{(4)^2 + (-3)^2 + (5)^2 + (8)^2 + (11)^2 + (-2)^2 + (0)^2 + (-5)^2}{8} - (2.250)^2 = 27.9375$$
$$s = \sqrt{27.9375} = 5.2856$$

Notice that the calculated quantities include fractional values, even though the data values are all integers.

4.11. Show that the two formulas for the variance, presented in Section 4.4, are equivalent.

$$v = \frac{(x_1 - \bar{x})^2 + (x_2 - \bar{x})^2 + \cdots + (x_n - \bar{x})^2}{n}$$

$$= \frac{(x_1^2 - 2x_1\bar{x} + \bar{x}^2) + (x_2^2 - 2x_2\bar{x} + \bar{x}^2) + \cdots + (x_n^2 - 2x_n\bar{x} + \bar{x}^2)}{n}$$

$$= \frac{(x_1^2 + x_2^2 + \cdots + x_n^2) - 2\bar{x}(x_1 + x_2 + \cdots + x_n) + n\bar{x}^2}{n}$$

$$= \frac{x_1^2 + x_2^2 + \cdots + x_n^2}{n} - 2\bar{x}^2 + \bar{x}^2$$

$$= \frac{x_1^2 + x_2^2 + \cdots + x_n^2}{n} - \bar{x}^2$$

4.12. Repeat the calculation in Example 4.14, using the rounded value for the mean, $\bar{x} = 11.32$ kg. Compare the answer with the result obtained in Example 4.14.

$$v = \frac{(11.33)^2 + (11.27)^2 + (11.38)^2 + (11.30)^2 + (11.29)^2 + (11.30)^2 + (11.34)^2 + (11.31)^2 + (11.32)^2}{9} - (11.32)^2$$

$$= -0.099\,69 \text{ kg}^2$$

The correct answer, obtained in Example 4.14 using $\bar{x} = 11.315\,556$ kg, is $v = 0.000\,903\,5$ kg^2. Thus we see that a very large error can result if we use a rounded value for $\bar{x}$ when calculating the variance.

4.13. Group the following data into the intervals 0–1.99, 2.00–3.99, 4.00–5.99, etc., and calculate the relative frequency for each interval. 6.05, 2.82, 14.27, 12.97, 7.36, 8.68, 12.53, 3.59, 10.82, 3.32, 10.59, 11.26, 10.28, 8.88, 1.24, 11.95, 13.28, 12.48, 8.60, 5.92, 8.18, 11.80, 9.87, 15.77, 7.46, 6.57, 8.03, 13.61, 10.93, 12.05, 15.62, 11.79, 9.03, 8.48, 5.89, 15.64, 9.61, 4.13, 15.18, 7.82, 7.97, 9.22, 8.91, 9.85, 9.31, 11.02, 11.15, 3.98, 17.79, 14.53, 16.05, 16.34, 1.74, 9.38, 13.15, 6.85, 12.69, 12.01, 8.88, 17.84.

     See Table 4-7.

Table 4-7

Interval	Number of Data Points	Relative Frequency, %
0.00–1.99	2	3.3
2.00–3.99	4	6.7
4.00–5.99	3	5.0
6.00–7.99	7	11.7
8.00–9.99	15	25.0
10.00–11.99	10	16.6
12.00–13.99	9	15.0
14.00–15.99	6	10.0
16.00–17.99	4	6.7
18.00–19.99	0	0
TOTALS	60	100.0%

4.14. Calculate the mean and the standard deviation for *(a)* the set of data in Problem 4.13, *(b)* the grouped data obtained therefrom in Table 4-7. Which pair of values is more accurate?

(a)
$$\bar{x} = \frac{6.05 + 2.82 + 14.27 + \cdots + 8.88 + 17.84}{60} = 10.050\,167 \approx 10.05$$

$$s = \left[\frac{(6.05)^2 + (2.82)^2 + (14.27)^2 + \cdots + (8.88)^2 + (17.84)^2}{60} - (10.050\,167)^2\right]^{1/2} = 3.90$$

(b)
$$\bar{x} = (0.033)(1.00) + (0.067)(3.00) + \cdots + (0.067)(17.00) = 9.9680 \approx 9.97$$

$$s = \{[(0.033)(1.00)^2 + (0.067)(3.00)^2 + \cdots + (0.067)(17.00)^2] - (9.9680)^2\}^{1/2} = 3.98$$

The values calculated in *(a)* are necessarily more accurate, though such individual data points are not always available.

4.15. Find the cumulative distribution of the grouped data presented in Table 4-7.

     See Table 4-8.

Table 4-8

Interval	Relative Frequency, %	Cumulative Relative Frequency, %
0.00–1.99	3.3	3.3
2.00–3.99	6.7	10.0
4.00–5.99	5.0	15.0
6.00–7.99	11.7	26.7
8.00–9.99	25.0	51.7
10.00–11.99	16.6	68.3
12.00–13.99	15.0	83.3
14.00–15.99	10.0	93.3
16.00–17.99	6.7	100.0

4.16. Determine the median for the cumulative distribution obtained in Table 4-8.

$$\tilde{x} = 8.00 + \frac{50-26.7}{51.7-26.7}(9.99 - 8.00) = 9.85$$

4.17. Fit a normal distribution to the cumulative distribution obtained in Table 4-8. Determine the mean and the standard deviation of the normal distribution.

Figure 4-12 shows a plot of the cumulative distribution data on probability paper. The data can be reasonably well represented by a straight line, though some scatter is present. This line represents the normal distribution function. From the plot it is seen that the normal distribution has a mean of 10.0 and a standard deviation of $10.0 - 6.0 = 4.0$. The equation for the normal density function (i.e. the symmetric, bell-shaped curve) is therefore

$$f(x) = \frac{1}{4\sqrt{2\pi}} e^{-[(x-10)/4]^2/2} = 0.099\,735\,6\,e^{-[(x-10)/4]^2/2}$$

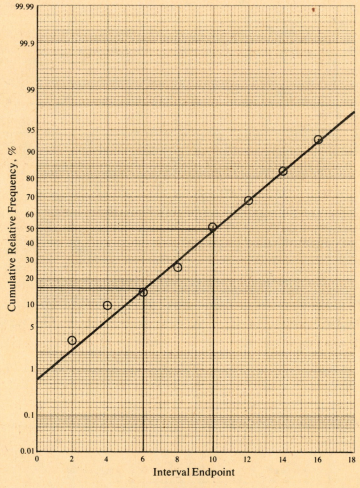

Fig. 4-12

4.18. Adjust the frequency distribution obtained in Table 4-7 so that the area under the frequency histogram is equal to unity. Plot the adjusted data on cartesian coordinates. Superimpose the normal density function obtained in Problem 4.17, and judge how well it represents the adjusted frequency distribution.

Each rectangle of the histogram has area given by

$$\text{area} = (\text{interval width}) \times (\text{height})$$

Therefore

$$\text{total area} = (1.99 - 0.00)(3.3) + (3.99 - 2.00)(6.7) + (5.99 - 4.00)(5.0) + \cdots + (17.99 - 16.00)(6.7)$$
$$= 199\%$$

Since we want to adjust the frequency distribution so that the total area is unity, we divide each of the relative frequencies in Table 4-7 by 199%. The adjusted relative frequencies are plotted against their corresponding interval midpoints in Fig. 4-13. The solid curve is the normal curve obtained in Problem 4.17, i.e. the graph of

$$f(x) = \frac{1}{4\sqrt{2\pi}} e^{-[(x-10)/4]^2/2}$$

The agreement between the solid curve and the data points is only fair, because of the data scatter.

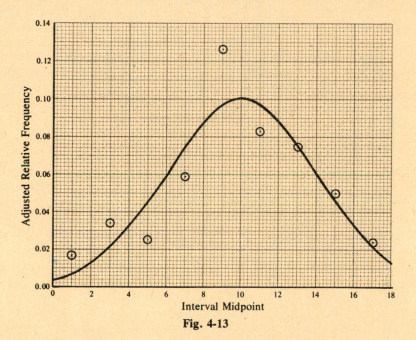

Fig. 4-13

4.19. Derive the least-squares equations for fitting a straight line to a set of M data points.

The equation for a straight line is $y = c_1 + c_2 x$. Therefore the sum of the squared errors can be written

$$S = \sum_{i=1}^{M} e_i^2 = \sum_{i=1}^{M} [y_i - (c_1 + c_2 x_i)]^2$$

We wish to determine the values of c_1 and c_2 that will cause S to be minimized. We therefore set the partial derivatives of S with respect to c_1 and c_2 equal to zero, i.e.

$$\frac{\partial S}{\partial c_1} = -2 \sum_{i=1}^{M} [y_i - (c_1 + c_2 x_i)] = 0$$

$$\frac{\partial S}{\partial c_2} = -2 \sum_{i=1}^{M} x_i [y_i - (c_1 + c_2 x_i)] = 0$$

These equations can be rearranged to read

$$\sum_{i=1}^{M} y_i - \sum_{i=1}^{M} c_1 - \left(\sum_{i=1}^{M} x_i\right) c_2 = 0$$

$$\sum_{i=1}^{M} x_i y_i - \left(\sum_{i=1}^{M} x_i\right) c_1 - \left(\sum_{i=1}^{M} x_i^2\right) c_2 = 0$$

or, rearranging further,

$$M c_1 + \left(\sum_{i=1}^{M} x_i\right) c_2 = \sum_{i=1}^{M} y_i$$

$$\left(\sum_{i=1}^{M} x_i\right) c_1 + \left(\sum_{i=1}^{M} x_i^2\right) c_2 = \sum_{i=1}^{M} x_i y_i$$

which are the desired final equations.

4.20. Use the method of least squares to pass a power function through the following 8 data points:

x	2	4	7	10	20	40	60	80
y	45	26	17	12.5	7.3	4.2	3.0	2.4

Following the procedure of Section 4.9, we seek the least-squares straight line

$$Y = c_1 + c_2 X$$

through the points $(\log x_i, \log y_i)$. Thus we form Table 4-9.

Table 4-9

x	$X=\log x$	X^2	y	$Y=\log y$	XY
2	0.301 030	0.090 619	45	1.653 212	0.497 666
4	0.602 060	0.362 476	26	1.414 973	0.851 899
7	0.845 098	0.714 191	17	1.230 449	1.039 850
10	1.000 000	1.000 000	12.5	1.096 910	1.096 910
20	1.301 030	1.692 679	7.3	0.863 323	1.123 209
40	1.602 060	2.566 596	4.2	0.623 249	0.998 482
60	1.778 151	3.161 822	3.0	0.477 121	0.848 393
80	1.903 090	3.621 751	2.4	0.380 211	0.723 576
TOTALS	9.332 519	13.210 134		7.739 448	7.179 985

The column sums of Table 4-9 provide the coefficients of the unknown constants c_1 and c_2 in the least-squares equations:

$$8c_1 + 9.332\,519 c_2 = 7.739\,448$$
$$9.332\,519 c_1 + 13.210\,134 c_2 = 7.179\,985$$

Solving, $c_1 = 1.895\,695$, $c_2 = -0.795\,724$. Then the desired power function is $y = ax^b$, with

$$a = \text{antilog } c_1 = 78.6493 \qquad b = c_2 = -0.795\,724$$

i.e.

$$y = 78.6 x^{-0.796}$$

This equation is plotted, along with the data points, in Fig. 4-14.
In Example 3.14 we fitted the equation

$$y = 80 x^{-0.801}$$

to this same set of data by simply plotting the data on a log-log graph and passing a straight line through the data points (see Fig. 3-18). This equation may be satisfactory for some purposes, but it will be less accurate than the power function obtained above using the least-squares procedure.

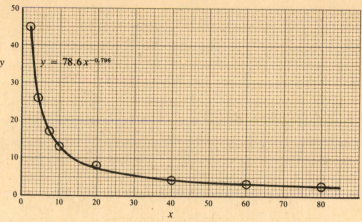

Fig. 4-14

4.21. The following data describe the growth of a bacteria culture. The population density (d), in parts per million, is tabulated as a function of time (t).   Determine an exponential equation that will represent the population density as a function of time, using the method of least squares.

t, days	0	0.5	1.0	1.5	2.0	2.5
d, ppm	0.10	0.45	2.01	9.00	40.34	180.80

Following the procedure of Section 4.9, we seek the least-squares straight line

$$D = c_1 + c_2 T$$

through the points $(t_i, \ln d_i)$.   Thus we form Table 4-10.

Table 4-10

$T = t$	T^2	d	$D = \ln d$	TD
0	0	0.10	−2.302 585	0
0.5	0.25	0.45	−0.798 508	−0.399 254
1.0	1.00	2.01	0.698 135	0.698 135
1.5	2.25	9.00	2.197 225	3.295 837
2.0	4.00	40.34	3.697 344	7.394 687
2.5	6.25	180.80	5.197 391	12.993 479
TOTALS 7.5	13.75		8.689 002	23.982 884

The least-squares equations for c_1 and c_2 are then:

$$6c_1 + 7.5c_2 = 8.689\,002$$
$$7.5c_1 + 13.75c_2 = 23.982\,884$$

Solving, $c_1 = -2.300\,871$ and $c_2 = 2.999\,230$.   The desired exponential function is

$$d = ae^{bt}$$

with

$$a = e^{c_1} = 0.100\,172 \qquad b = c_2 = 2.999\,230$$

that is, $d = 0.100\,e^{3.00\,t}$,   which is essentially the same result obtained by plotting the data on a semilog graph and passing a straight line through the data points (see Problem 3.2).

The exponential function is shown plotted, along with the data points, in Fig. 4-15.

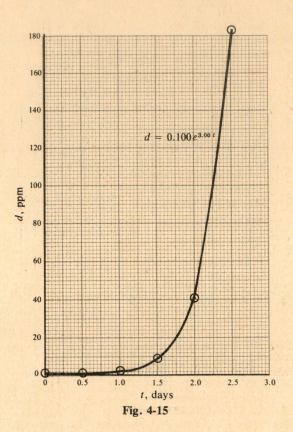

$$d = 0.100\,e^{3.00\,t}$$

Fig. 4-15

4.22. Pass a quadratic curve through the following set of data, using the method of least squares.

x	0	0.5	1.0	1.5	2.0	2.5	3.0	3.5	4.0	4.5
y	1.8	3.2	5.5	8.0	12.5	15.9	22.6	28.7	35.5	48.4

The quadratic equation that will be fitted to the data points is

$$y = c_1 + c_2 x + c_3 x^2$$

The least-squares equations for c_1, c_2, and c_3 follow from Section 4.9, for the case $n = 2$. Thus:

$$Mc_1 + (\Sigma x_i)c_2 + (\Sigma x_i^2)c_3 = \Sigma y_i$$
$$(\Sigma x_i)c_1 + (\Sigma x_i^2)c_2 + (\Sigma x_i^3)c_3 = \Sigma x_i y_i$$
$$(\Sigma x_i^2)c_1 + (\Sigma x_i^3)c_2 + (\Sigma x_i^4)c_3 = \Sigma x_i^2 y_i$$

Here, $M = 10$, and the remaining coefficients are calculated as in Table 4-11.

The least-squares equations can now be written as

$$10c_1 + \quad 22.5c_2 + \quad 71.25c_3 = 182.1$$
$$22.5c_1 + \quad 71.25c_2 + \quad 253.125c_3 = 611.9$$
$$71.25c_1 + 253.125c_2 + 958.3125c_3 = 2276.75$$

The solution to these three simultaneous equations is

$$c_1 = 2.472\,73 \qquad c_2 = 0.325\,15 \qquad c_3 = 2.106\,06$$

Therefore the desired quadratic equation is

$$y = 2.472\,73 + 0.325\,15x + 2.106\,06x^2$$

Table 4-11

x	x^2	x^3	x^4	y	xy	x^2y
0	0	0	0	1.8	0	0
0.5	0.25	0.125	0.0625	3.2	1.6	0.8
1.0	1.00	1.000	1.0000	5.5	5.5	5.5
1.5	2.25	3.375	5.0625	8.0	12.0	18.0
2.0	4.00	8.000	16.0000	12.5	25.0	50.0
2.5	6.25	15.625	39.0625	15.9	39.75	99.375
3.0	9.00	27.000	81.0000	22.6	67.80	203.400
3.5	12.25	42.875	150.0625	28.7	100.45	351.575
4.0	16.00	64.000	256.0000	35.5	142.00	568.000
4.5	20.25	91.125	410.0625	48.4	217.80	980.100
TOTALS 22.5	71.25	253.125	958.3125	182.1	611.90	2276.750

4.23. Show that the correlation coefficient for a set of data points (x_i, y_i) is independent of the units of measurement of the x_i and the y_i.

If the units of x_i are changed (e.g. from feet to meters), then x_i is replaced by

$$x_i' = px_i$$

where p is the appropriate units conversion factor (e.g. 0.3048 m/ft). Likewise, $y_i' = qy_i$. The correlation coefficient of the data points (x_i', y_i') is then

$$r' = \frac{M \sum x_i' y_i' - (\sum x_i')(\sum y_i')}{\{[M \sum x_i'^2 - (\sum x_i')^2][M \sum y_i'^2 - (\sum y_i')^2]\}^{1/2}}$$

$$= \frac{Mpq \sum x_i y_i - (p \sum x_i)(q \sum y_i)}{\{[Mp^2 \sum x_i^2 - (p \sum x_i)^2][Mq^2 \sum y_i^2 - (q \sum y_i)^2]\}^{1/2}}$$

$$= \frac{pq}{pq} \frac{M \sum x_i y_i - (\sum x_i)(\sum y_i)}{\{[M \sum x_i^2 - (x_i)^2][M \sum y_i^2 - (\sum y_i)^2]\}^{1/2}} = r$$

where r is the correlation coefficient of the data points (x_i, y_i).

It can be shown that the above result holds even when the change in units involves a change in zero point (e.g. conversion from °F to °C).

Supplementary Problems

Answers to most problems are provided at the end of the book.

4.24. A voltmeter has a scale that ranges from 0 to 500 V.   The scale is divided into 10 V intervals.   (*a*) If the position of the needle can be determined to the nearest half interval, what will be the uncertainty associated with the voltmeter when using this scale?   (*b*) Suppose that the needle is located approximately midway between the 170 and the 180 V divisions.   What voltage reading should be accepted, if the uncertainty in the measurement is to be confined to the last significant figure?   (*c*) How should the measurement in (*b*) be recorded if the uncertainty in the measurement is to be specified along with the measured value?   (*d*) What voltage should be recorded if the needle is located approximately at the 340 V division?   (*e*) What voltage should be recorded if the needle is located near the 60 V division?

4.25. A flowmeter has a scale that ranges from from 0 to 100 m³/min. The scale is divided into intervals of 5 m³/min. (*a*) Suppose that the position of the needle can be determined to the nearest half interval. What will be the uncertainty associated with the flowmeter? (*b*) What flow rate should be recorded if the needle is located near the 65 m³/min division? (*c*) What flow rate should be recorded if the needle is located approximately midway between the 20 and the 25 m³/min divisions? (*d*) What flow rate should be recorded if the needle is located approximately midway between the 75 and 80 m³/min divisions?

4.26. The following data indicate the boiling point (b.p.) of an alcohol-water mixture as a function of alcohol concentration (*x*).

x, %	1	2	4	6	8	10	15	20	25	30	40	50
b.p., °C	102	99.5	97.5	96.5	94	92.5	90.5	89.5	88	86	85.5	83.5

Plot the data and determine (*a*) if consistent errors are present in the data, (*b*) if random errors are present, (*c*) how more accurate values may be obtained if the data do appear to contain errors.

4.27. Several measurements have been made of the freezing point of a 5% salt solution, yielding the values −3.08, −2.90, −2.94, −3.05, −2.97, −3.14, and −3.02 °C. (*a*) What is the average value obtained for the freezing point? Does this value coincide with any of the individual measurements? (*b*) Determine the deviations about the average, to two-decimal accuracy. (*c*) Suppose that those measurements whose deviations exceed 0.1 °C in magnitude are rejected as being inaccurate. How will this affect the average value of the freezing point?

4.28. A student has conducted an experiment to determine how far a spring must be extended in order that the restoring force be 10 N. The following measurements were obtained: 12.2, 11.7, 12.4, 12.0, 10.4, 11.9, and 12.4 cm. (*a*) What is the average value obtained for these measurements? (*b*) Calculate the deviations about the average. (*c*) Suppose that those measurements whose deviations exceed 1 cm in magnitude are rejected as being inaccurate. How will this affect the average? (*d*) Recalculate the deviations of the remaining measurements about the new average.

4.29. A student has determined that the earth's gravitational acceleration is 9.819 m/s² (at sea level and a latitude of 45°). The correct value is known to be 9.807 m/s². Determine (*a*) the absolute error, (*b*) the relative error, (*c*) the percent error.

4.30. A motorcyclist has determined that his speedometer reads 3% too high. Determine the correct speeds corresponding to speedometer readings of (*a*) 30 mph, (*b*) 55 mph, (*c*) 80 mph.

4.31. An engineering student purchased a 1 lb box of candy for his girlfriend, thinking that the candy itself weighed 1 lb. Before presenting his girlfriend with the candy the student determined that the sealed box weighed 465 grams (note that the scale was calibrated in mass units). He later weighed the empty box and the wrappings, and found that they weighed 42 grams. (*a*) What is the percent error in the weight of the candy? (*b*) If the box of candy cost $6.50, by how much was the student overcharged?

4.32. A capacitor has a nominal rating of 500 μF. (*a*) If the actual capacitance is 522 μF, what is the percent error, based upon the rated capacitance? (*b*) If the accuracy in the rated capacitance is ±10%, in what range can the actual capacitance be expected to fall?

4.33. A wooden meterstick was used to measure the dimensions of a rectangular tank. The measured dimensions were 1.275 × 1.886 × 6.351 "meters." The true length of the meterstick was subsequently found to be only 99.35 centimeters. (*a*) What is the actual volume of the tank? (*b*) What percent error is associated with the volume that is calculated from the original measurements?

4.34. A motorist has been checking his automobile's gasoline mileage by recording the odometer reading and the amount of gasoline required to fill his tank each time he stops for gas. The following data were obtained during the month of July.

Odometer Reading	Gasoline Purchased
18 258 mi	16.4 gal
18 673	18.7
19 059	16.0
19 427	17.2
19 881	18.3
20 244	15.8
20 654	17.9

(*a*) What is the overall gasoline mileage for the entire month, based upon the above figures? (*b*) The motorist suspects that his odometer is reading 5% too high. If this is so, what will be the true gasoline mileage for the entire month?

4.35. Four measurements have been obtained for the voltage drop across a resistor, resulting in an average value of 261.7 V. Subsequently, two additional measurements, whose values are 257.4 V and 259.0 V, have been obtained. Determine an average value for all six measurements.

4.36. Two different types of coal are to be blended together. The first type contains 6.4% sulfur, and the second type contains 2.8% sulfur. (*a*) What will be the sulfur content of a blended product consisting of 70% high-sulfur coal and 30% low-sulfur coal? (*b*) In what quantities must the two types of coals be blended if the sulfur content of the blended product is to be 4.5%?

4.37. Two different groups of students have been conducting experiments to determine the atmospheric pressure at an altitude of 8000 ft above sea level. The first group obtained an average value of 557.8 mmHg and the second group determined an average value of 562.6 mmHg. It was decided that the results of the first group would be given twice as much emphasis as the results of the second group. What final value should be accepted for the pressure?

4.38. An engineering student has taken the following courses and earned the following grades during the past term. Determine the student's QPA (Problem 4.9), assuming that A = 4, B = 3, etc.

Course	Credit Hours	Grade
Linear Algebra	3	B
Physics II	4	A
Economics	3	C
Solid Mechanics	3	B
Electronics	3	C
Electronics Lab.	1	A

4.39. At the end of her freshman year an engineering student has taken 34 credit hours and has earned an overall QPA (covering all freshman courses) of 3.41. During the first term of her sophomore year she took 16 credit hours, for which she earned a QPA of 3.23. What is her overall QPA?

4.40. A student's four-year academic record is shown below on a term-by-term basis. Determine the student's overall QPA.

Term	1	2	3	4	5	6	7	8
Credit Hours	16	17	16	15	17	16	18	15
QPA	2.41	2.76	3.22	3.08	2.94	3.35	3.20	2.83

4.41. Determine the mean, the median, the mode(s), the variance, and the standard deviation for each of the following sets of data: (a) 8.2, 8.5, 7.6, 8.0, 7.4, 8.2, 7.8; (b) 137, 128, 117, 144, 160, 132, 155, 141; (c) 65.4, 64.7, 65.4, 67.2, 63.8, 66.9, 67.2, 66.1, 65.0; (d) −4.2, 5.9, 2.0, −3.6, −1.1, 3.9, 0.8.

4.42. Two groups of students have been measuring the yield point of a metal alloy. The following data have been obtained by each group: 3865, 3407, 3620, 3883, 3571 N (group A) and 3462, 3775, 3208, 3550, 3919, 3367, 3889 N (group B). Which set of data appears to be more accurate, and why?

4.43. A small city has six firms which employ engineers in sizable numbers. The mean salary for each firm is given below. Determine the overall (city-wide) mean salary.

Firm	Mean Engineers' Salary	Number of Engineers Employed
A	$24 855	85
B	27 324	43
C	25 590	68
D	21 400	177
E	26 718	24
F	28 911	16

4.44. Four different groups of students have been measuring the resistance of an electrical device. The following results have been obtained.

Group	1	2	3	4
Avg. Resistance, Ω	77.2	67.7	74.6	78.0
No. of Measurements	12	15		8

If an overall mean value of 73.6 Ω was found, how many measurements were taken by group 3?

4.45. The set of data below represents the final exam scores for a class of 50 engineering students. Group the scores into intervals of 5 percentage points, i.e. 0.01–5, 5.01–10, 10.01–15, ..., 95.01–100. Calculate the relative frequency for each interval.

47	78	89	99	62	23	47	76	48	97
56	44	83	12	92	76	53	77	56	86
56	20	94	79	67	48	29	52	70	69
41	100	66	68	43	71	68	85	63	10
38	46	82	100	61	79	90	76	61	44

4.46. Calculate the mean and the standard deviation for (a) the individual exam scores and (b) the grouped data, in Problem 4.45.

4.47. Calculate a cumulative distribution from the grouped data obtained in Problem 4.45.

4.48. Determine the median value for the cumulative distribution obtained in Problem 4.47.

4.49. Determine the mean, the median, and the standard deviation for the data presented in Problem 3.4.

4.50. Determine the mean, the median, and the standard deviation for the data presented in Table 3-3.

4.51. Suppose that the distribution of final exam scores given in Problem 4.45 is representative of the performance of freshman engineering students from one year to the next. In a class of 100 students, how many students would be expected to have final exam scores (a) less than or equal to 35? (b) greater than 80? (c) greater than 60 but not greater than 75? (*Hint*: Use the cumulative distribution data obtained in Problem 4.47.)

4.52. Figure 4-16 presents a cumulative distribution curve showing the probability that a plastic material contains various levels of impurity. If this distribution curve is representative of the manufacturing process, find the probability that the impurity level of a given batch of plastic is (a) not more than 2.5 ppm, (b) greater than 2.5 ppm, (c) not more than 5 ppm, (d) greater than 5 ppm, (e) between 2.5 and 5 ppm, (f) between 3.5 and 4 ppm.

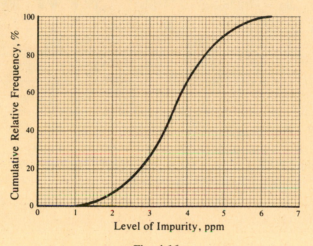

Fig. 4-16

4.53. From Fig. 4-16, (a) what minimum level of impurity can always be achieved with this manufacturing process? (b) what level of impurity is virtually unachievable with this manufacturing process?

4.54. From the data presented in Table 3-3, what is the probability that a cylinder in an engine block will have a diameter (a) between 89.7 and 89.9 mm? (b) between 90.1 and 90.5 mm? (c) less than or equal to 89.7 mm? (d) greater than 89.7 mm? (e) greater than 90.7 mm?

4.55. Determine the mean and the standard deviation for the cumulative normal distributions shown in (a) Fig. 3-25, (b) Fig. 3-35. (Obtain the desired values from the straight line, not the actual data points.)

4.56. A normal distribution has a mean of 20 and a standard deviation of 5. Plot (a) the normal curve on arithmetic coordinates, using the formula for $f(x)$ given in Section 4.8 or using tabulated values of $f(x)$ from a handbook; (b) the cumulative distribution on probability paper. Which curve better conveys the basic character of the normal distribution? Which curve is easier to construct and to read?

4.57. Use probability paper to fit a normal distribution to the cumulative distribution data obtained in Problem 4.47. What are the mean and the standard deviation for the normal distribution? Compare with the values obtained from the grouped data [see Problem 4.46(b)].

4.58. Determine the median and the mode for the normal distribution obtained in Problem 4.57.

4.59. What would happen to a normal curve if (*a*) the mean were increased or decreased? (*b*) the standard deviation were increased or decreased?

4.60. Fit a straight line to the data presented in Problem 3.1, using the method of least squares.

4.61. The following data indicate the sulfur content (*S*) of a coal sample as a function of its density (ρ). Fit a straight line to the data, using the method of least squares.

ρ, 10^3 kg/m³	1.30	1.35	1.40	1.45	1.50	1.60	1.80	2.00	2.20
S, %	0.85	1.02	1.23	1.47	1.77	2.58	3.46	4.40	5.51

4.62. Fit a straight line to the following set of data, using the method of least squares.

x	−0.03	−0.02	−0.01	0	0.01	0.02	0.03	0.04
y	6.2	4.8	2.5	1.0	−1.2	−2.7	−4.8	−6.9

4.63. Fit power functions to the following sets of data, using the method of least squares: (*a*) Problem 3.3, (*b*) Problem 3.29.

4.64. Fit exponential functions to the following sets of data, using the method of least squares: (*a*) Example 3.11, (*b*) Problem 3.2, (*c*) Problem 3.30.

4.65. The following data represent the resistance (*R*) of a rheostat (a variable resistor) as a function of the dial setting (*x*).

x	0	0.1	0.3	0.6	1.3	2.8	5.4	8.7	15.0	28.8
R, Ω	0	10	20	50	100	200	350	500	700	900

Fit an appropriate equation to the data using the method of least squares. (*Hint*: Plot $1000 - R$ against *x*.)

4.66. ABC Electronics is a young, rapidly growing company whose annual earnings (per share of common stock) are given below.

Year	1	2	3	4	5	6	7	8	9	10
Earnings, $/share	0.01	0.02	0.02	0.03	0.03	0.04	0.06	0.09	0.24	0.38

Year	11	12	13	14	15	16	17	18	19	
Earnings	0.63	0.93	1.24	1.48	1.73	2.07	2.50	3.12	3.48	

Use the method of least squares to fit a curve through the above data, choosing (*a*) a power function, (*b*) a quadratic function. (*c*) Which curve provides the better fit?

4.67. For the data of Problem 4.62 calculate the means $\bar{x}$ and $\bar{y}$, and show that the least-squares straight line for the data passes through the point $(\bar{x}, \bar{y})$. (This is true for any least-squares line.)

4.68.   Calculate the correlation coefficient for each of the following sets of data.   What conclusion can be drawn in each case?

(a)

x	0.5	1.0	1.5	2.0	2.5	3.0	3.5	4.0	4.5	5.0
y	1.3	1.8	2.0	2.1	2.4	2.7	2.8	3.0	3.4	3.6

(b)

x	0.5	1.0	1.5	2.0	2.5	3.0	3.5	4.0	4.5	5.0
y	1.7	4.3	2.4	3.6	1.1	2.7	3.8	1.5	2.9	0.8

(c)

x	0	0.5	1.0	1.5	2.0	2.5	3.0	3.5	4.0	4.5	5.0
y	3.6	3.0	2.6	2.3	1.8	1.3	0.7	0.2	-0.2	-0.5	-1.3

Chapter 5

Approximate Solution of Equations

Of the many tasks confronting the engineer, none occurs more frequently than the need to solve various kinds of mathematical equations.   In some situations a computer may be required, in order to obtain a solution as accurately as possible.   This is particularly true of detailed design problems, or of problems in automatic control.   More often, however, an approximate solution, to two or three significant figures, will be adequate.   Under these circumstances the sophistication of a computer solution is not only unnecessary, but may even be undesirable.   What is needed is the ability to obtain approximate solutions quickly and easily, using nothing more than a pocket calculator and, perhaps, a sheet of graph paper.

5.1   GRAPHICAL SOLUTION OF ALGEBRAIC EQUATIONS

An algebraic equation is said to be *linear* if it contains the unknown quantity raised only to the first power; for example,

$$5x - 6 = 14$$

Such equations are easily solved by direct algebraic rearrangement.

Many of the equations that arise in science and engineering are *nonlinear*, in that they include terms in which the unknown quantity is raised to powers other than one, or involve special functions of the unknown quantity.   Most nonlinear equations, e.g.

$$x^2 = 10 \cos x$$

cannot be solved by simple algebraic rearrangement.

An easy way to solve a nonlinear equation is to rearrange the equation into the form $f(x) = 0$ and plot $f(x)$ against x.   A point where $f(x)$ intersects the abscissa will be a value of x that causes $f(x)$ to equal zero.   Hence this value of x will be a solution (more precisely, a *real root*) of the equation.   The method results in solutions that are accurate to two or three significant figures. This is adequate for many engineering purposes.

Example 5.1.   Find a positive root of the equation $x^2 = 10 \cos x$.
We first rearrange the equation into the form $f(x) = 0$.   Thus,

$$f(x) = x^2 - 10 \cos x = 0$$

We then construct a table of $f(x)$ versus x, as shown below.

x	0	0.5	1.0	1.5	2.0
$f(x)$	-10	-8.526	-4.403	1.543	8.161

These calculations reveal that there is a positive root somewhere between $x = 1.0$ and $x = 1.5$.   We therefore tabulate a few more values within this interval, as shown below.

x	1.0	1.1	1.2	1.3	1.4	1.5
$f(x)$	-4.403	-3.326	-2.184	-0.985	0.260	1.543

We now plot $f(x)$ against x as shown in Fig. 5-1.   From this figure we see that $f(x)$ crosses the x-axis at about $x = 1.38$.   Thus $x = 1.38$ is the desired root.

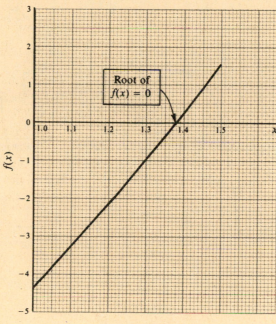

Fig. 5-1

Many nonlinear equations have more than one real root.   In fact, certain equations involving trigonometric terms have an *infinite* number of real roots.

Example 5.2.   The equation $x \cos x = 0$ has an infinite number of real roots in the interval $x \geq 0$, since $\cos x = 0$ when $x = \pi/2, 3\pi/2, 5\pi/2, \ldots$.   The left-hand side will also equal zero when $x = 0$.

On the other hand, some nonlinear equations have no real roots at all (i.e. all the roots will be imaginary or complex).   It is quite common for a nonlinear equation to have both real and complex (or imaginary) roots.

Example 5.3.   The polynomial equation $x^3 - 3x^2 + 10x - 30 = 0$ can be factored into the form

$$(x - 3)(x^2 + 10) = 0$$

Hence its roots are $x = 3$, $x = \sqrt{10}\,i$, and $x = -\sqrt{10}\,i$. Thus one root is real and two are imaginary.

The above graphical procedure can be used only to find real roots.   This is usually not a problem, however, since most elementary technical calculations are concerned only with the real roots of an equation.

It is helpful to recognize situations that allow more than one real root.   For a polynomial equation, the total number of roots, real and complex, is equal to the highest power of the unknown quantity.   Complex roots must occur in pairs.   Hence, if the highest power is odd, the equation possesses an odd number of real roots (hence, at least one real root).

Example 5.4.   The equation $f(x) = x^5 - 3x^4 - 3x^2 - 35x + 60 = 0$ has at least one real root, since the highest power of x is odd.   Figure 5-2 shows a plot of $f(x)$ against x for the interval $-4 \leq x \leq 4$.   (The plot was obtained by constructing a table of $f(x)$ versus x, as in Example 5.1.)   From this figure we see that the equation has three real roots within the given interval; namely, $x = -2.2$, $x = 1.4$, and $x = 3.6$.   It is not hard to see that the equation has no other real roots.

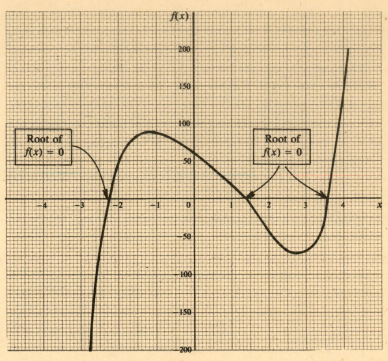

Fig. 5-2

5.2 NUMERICAL SOLUTION OF ALGEBRAIC EQUATIONS

Nonlinear algebraic equations can be solved to a very high degree of accuracy by means of various computational procedures which have been devised for this purpose. These methods are all based upon the use of repeated calculations to successively refine earlier results. Each successive calculation is called an *iteration*. Hence these methods are sometimes referred to as *iterative techniques*. Such methods can be very useful when the solutions obtained by graphical procedures are not sufficiently accurate.

Successive Substitution

The method of successive substitution begins with an approximate solution to an algebraic equation, and gradually refines that solution through a sequence of trial-and-error-type calculations.

In order to use this method we must first put the given equation into the form

$$x = g(x)$$

We then *guess* a value for the desired root (an approximate, graphical solution will serve very well). Let us call this value x_0. We then substitute x_0 into the right side of the equation and compute a new value of x, call it x_1. In other words,

$$x_1 = g(x_0)$$

This procedure is shown graphically in Fig. 5-3(a).

If the value chosen for x_0 happened to be the exact value of the root, then x_1 would equal x_0, and the computation would cease. It is most likely, however, that x_0 will differ somewhat from the true root, so that x_1 will not equal x_0. We then substitute x_1 into the right side of the equation and again calculate a new value for x, i.e.

$$x_2 = g(x_1)$$

as shown in Fig. 5-3(b).

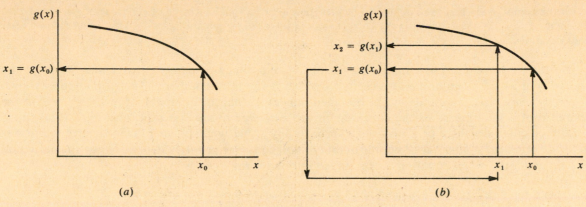

Fig. 5-3

This procedure can be repeated over and over, using a computer or an electronic calculator, until we obtain two successive values of x that are sufficiently close together.

In general, we can write $x_i = g(x_{i-1})$, where x_{i-1} represents the currently assumed value of x and x_i represents the newly calculated value. Each successive calculation represents one iteration, as illustrated in Fig. 5-4.

The computation will terminate when the difference between x_{i-1} and x_i (disregarding sign) is less than some specified small value. Mathematically, the stopping criterion can be expressed as

$$|x_i - x_{i-1}| \leq \epsilon$$

where ϵ represents a preestablished, small positive number. The graphical interpretation of this criterion is shown in Fig. 5-4.

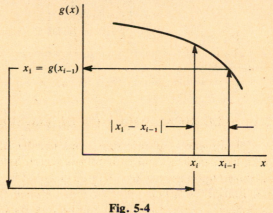

Fig. 5-4

Example 5.5. Solve the equation $x^2 = 10 \cos x$ for a positive real root. Use the method of successive substitution, with $\epsilon = 0.001$.

We first rearrange the equation to read

$$x = \cos^{-1}(0.1x^2)$$

(See Example 5.6 for another possibility.) We have already determined that the root is approximately 1.4 (see Example 5.1). Hence we can choose $x_0 = 1.4$, and we can write

$$x_1 = \cos^{-1}[(0.1)(1.4)^2] = 1.373\,519$$

Applying the stopping criterion, we see that

$$|x_1 - x_0| = |1.373\,519 - 1.4| = 0.026\,481$$

which is larger than ϵ. Hence we must perform additional iterations, as shown below

$$x_2 = \cos^{-1}[(0.1)(1.373\,519)^2] = 1.381\,003 \qquad |x_2 - x_1| = 0.007\,484$$
$$x_3 = \cos^{-1}[(0.1)(1.381\,003)^2] = 1.378\,904 \qquad |x_3 - x_2| = 0.002\,099$$
$$x_4 = \cos^{-1}[(0.1)(1.378\,904)^2] = 1.379\,494 \qquad |x_4 - x_3| = 0.000\,590$$

Since

$$|x_4 - x_3| < 0.001$$

we have satisfied the stopping criterion, and we accept as the final solution the value $x = 1.379\,494$.

If the computation had been continued further (because of a smaller choice of ϵ, perhaps), we would have obtained $x = 1.379\,365$, which is the correct answer to seven significant figures. This result follows after eight iterations.

So far we have been assuming that the calculated values of x move closer and closer to the desired root as the computation proceeds from one iteration to another. This condition is known as *convergence*. It cannot always be assumed, however, that the computation will converge to the desired root. In some cases the computation will converge to another root, or it may "blow up," i.e. the calculated values of x may not approach any value whatsoever. This latter condition is called *divergence*. Whether or not the iterations will converge properly depends upon the particular way in which the given equation is rearranged into the form $x = g(x)$.

Example 5.6. Rearrange the equation $x^2 = 10 \cos x$ into the form
$$x = \sqrt{10 \cos x}$$
Then solve for a positive real root, using the method of successive substitution with $\epsilon = 0.001$ (see Example 5.5).

If we again begin with $x_0 = 1.4$, we obtain the following sequence of calculations.

$$x_1 = \sqrt{10 \cos 1.4} = 1.303\,714 \qquad |x_1 - x_0| = 0.096\,286$$
$$x_2 = \sqrt{10 \cos 1.303\,714} = 1.624\,555 \qquad |x_2 - x_1| = 0.320\,841$$
$$x_3 = \sqrt{10 \cos 1.624\,555} = \sqrt{-0.537\,327}$$

The computation cannot be continued further without involving imaginary numbers, which is beyond our present scope of interest. We therefore terminate the computation at this point. It should be understood, however, that the computation would not necessarily converge to the correct answer, even if complex arithmetic were considered.

Many beginning students would be tempted to ignore the minus sign in the last calculation and simply continue with the computation. If this is done (and subsequent minus signs are also ignored), the computation does converge to $x = 3.161\,945$, after 7 iterations. However, this value is *not* a root of the given equation (it is a root of $x^2 = 10\,|\cos x|$).

How can we tell if the method when applied to a particular form, $x = g(x)$, will converge to a desired root, provided we begin with a value of x_0 which is reasonably close to the root? The answer to this question depends upon $g'(x)$, the first derivative of the function $g(x)$. In fact, the method will converge if and only if the magnitude of $g'(x)$ is less than 1 in the vicinity of the desired root. Mathematically, we can write the convergence criterion as
$$|g'(x)| < 1$$
(see Problem 5.5 for a derivation of this criterion).

Example 5.7. Apply the convergence criterion
$$|g'(x)| < 1$$
to the equation $x^2 = 10 \cos x$. Compare with the results obtained in Examples 5.5 and 5.6.

In Example 5.5 we had written $x = \cos^{-1}(0.1x^2)$. Thus
$$g(x) = \cos^{-1}(0.1x^2) \qquad \text{and} \qquad g'(x) = \frac{d}{dx}[\cos^{-1}(0.1x^2)] = -\frac{0.2\,x}{\sqrt{1 - 0.01x^4}}$$
Since the root has the value $x = 1.379\,365$ (see Example 5.5), we obtain
$$|g'(x)| = \frac{(0.2)(1.379\,365)}{\sqrt{1 - (0.01)(1.379\,365)^4}} = 0.28$$
Therefore $|g'(x)| < 1$, so that the computation should converge. This is consistent with the results obtained in Example 5.5.

In Example 5.6 we had $x = \sqrt{10 \cos x}$, so that
$$g(x) = \sqrt{10 \cos x} \qquad \text{and} \qquad g'(x) = \frac{d}{dx}\sqrt{10 \cos x} = -\frac{5 \sin x}{\sqrt{10 \cos x}}$$

If we evaluate this function at $x = 1.379\,365$, we obtain

$$|g'(x)| = \frac{5 \sin 1.379\,365}{\sqrt{10 \cos 1.379\,365}} = 3.56$$

Since $|g'(x)| > 1$, we see that the computation will **not converge** in this case.

If $g(x)$ is plotted against x, there is a simple graphical interpretation of the convergence criterion—namely, that in the vicinity of the desired root, *the magnitude of the slope of the tangent to $g(x)$ must be less than unity*.

The Newton-Raphson Method

The *Newton-Raphson method* (also known as *Newton's method*) is simply a special case of the method of successive substitution which, for many problems, converges very rapidly. In this method the function $g(x)$ is taken as

$$g(x) = x - \frac{f(x)}{f'(x)}$$

where $f(x) = 0$ represents the original algebraic equation. Then the iterative equation, $x_i = g(x_{i-1})$, becomes

$$x_i = x_{i-1} - \frac{f(x_{i-1})}{f'(x_{i-1})}$$

Example 5.8. Solve the equation $x^2 = 10 \cos x$ for a positive real root using the Newton-Raphson method. Choose $\epsilon = 0.001$, as in Example 5.5.

We first write

$$f(x) = x^2 - 10 \cos x \qquad \text{and} \qquad f'(x) = 2x + 10 \sin x$$

Therefore

$$g(x) = x - \frac{f(x)}{f'(x)} = x - \frac{x^2 - 10 \cos x}{2x + 10 \sin x} = \frac{x^2 + 10(x \sin x + \cos x)}{2x + 10 \sin x}$$

Let us begin with $x_0 = 1.4$, as in Example 5.5. Substituting into our expression for $g(x)$, we obtain

$$x_1 = \frac{(1.4)^2 + 10(1.4 \sin 1.4 + \cos 1.4)}{(2)(1.4) + 10 \sin 1.4} = 1.379\,428$$

Applying the stopping criterion, we see that

$$|x_1 - x_0| = |1.379\,428 - 1.4| = 0.020\,572$$

which is larger than ϵ. We therefore carry out an additional iteration.

$$x_2 = \frac{(1.379\,428)^2 + 10(1.379\,428 \sin 1.379\,428 + \cos 1.379\,428)}{(2)(1.379\,428) + 10 \sin 1.379\,428} = 1.379\,365 \qquad |x_2 - x_1| = 0.000\,063$$

At this point we have not only satisfied the stopping criterion, but we have obtained the exact answer, $x = 1.379\,365$, to seven significant figures. Only two iterations were required to obtain this result, whereas eight iterations were required in Example 5.5. The Newton-Raphson method converges very rapidly for this problem.

The Newton-Raphson method cannot be used to solve every algebraic equation; there are some equations for which it will not converge. In particular, the method tends to "blow up" when the value for $f'(x)$ is close to zero (see Problem 5.6). Nevertheless, use of the Newton-Raphson method is generally recommended when a converging type of iterative procedure is desired. The method is especially well suited for computer implementation. (We will discuss this last point in greater detail in Chapter 7.)

The Method of Bisection (Interval Halving)

We will now consider a basically different approach to the problem of finding a real root of an algebraic equation. Instead of attempting to calculate a value for the root itself, let us successively reduce the size of the interval that contains the desired root. Once we have obtained a sufficiently small interval, then the root can safely be assumed to lie at any convenient location within the interval.

Let us begin with some interval, $a \leq x \leq b$, which is known to contain one and only one real root of the equation $f(x) = 0$. Therefore the function $f(x)$ will cross the x-axis once and only once within the interval, as shown in Fig. 5-5. This intersection point will be the desired root.

We can find a smaller interval that contains the desired root by the following simple procedure. Evaluate $f(x)$ at the *midpoint* of the original interval, i.e. calculate $f(x_m)$, where

$$x_m = \frac{a + b}{2}$$

If the root is contained in the left half-interval ($a \leq x \leq x_m$), then $f(a)$ and $f(x_m)$ will have different signs, but $f(x_m)$ and $f(b)$ will have the same sign. On the other hand, if the root lies in the right half-interval ($x_m \leq x \leq b$), then $f(a)$ and $f(x_m)$ will have the same sign, but $f(x_m)$ and $f(b)$ will have different signs. Thus, by comparing the signs of $f(a)$, $f(x_m)$ and $f(b)$, one of the half-intervals can be retained, and the other rejected. The computation can then be repeated with the new half-interval.

The procedure is illustrated in Fig. 5-6, where two successive iterations are carried out. First the left half-interval is retained, as shown in Figs. 5-6(a) and (b), then the right half-interval, as shown in Figs. 5-6(b) and (c). Moreover, Fig. 5-6(c) reveals that the right half-interval will be retained in the next iteration.

Notice that we continue to refer to the left boundary as a, the right boundary as b, and the midpoint as x_m in each new interval. Therefore the point labeled x_m within a given iteration will become either point a or point b in the next iteration.

The elimination is continued until an interval is obtained whose size is less than some specified small value. In other words, the computation ceases once the condition

$$|b - a| < \epsilon$$

has been satisfied. The final solution will then be the value of x (i.e. a, b, or x_m) for which the corresponding value of $f(x)$ is smallest in magnitude.

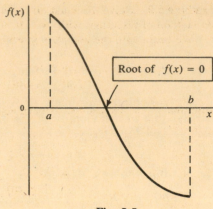

Fig. 5-5

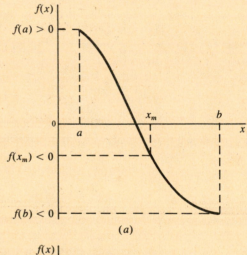

(a)

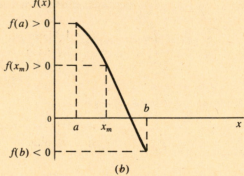

(b)

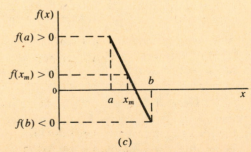

(c)

Fig. 5-6

Example 5.9. Solve the equation $x^2 = 10 \cos x$ for a positive real root within the interval $1 \leqslant x \leqslant 2$. Use the method of bisection, with $\epsilon = 0.001$.

We begin with the complete interval $1 \leqslant x \leqslant 2$. Hence, for the **1st iteration**,

$$a = 1 \qquad\qquad x_m = 1.5 \qquad\qquad b = 2$$
$$f(a) = -4.403 \qquad f(x_m) = 1.543 \qquad f(b) = 8.161$$

Note that $|b-a|$ exceeds ϵ. Hence additional iterations are required, starting with $1 \leqslant x \leqslant 1.5$.

2nd iteration

$$a = 1 \qquad\qquad x_m = 1.25 \qquad\qquad b = 1.5$$
$$f(a) = -4.403 \qquad f(x_m) = -1.591 \qquad f(b) = 1.543$$
$$|b - a| = 0.5$$

3rd iteration

$$a = 1.25 \qquad\qquad x_m = 1.375 \qquad\qquad b = 1.5$$
$$f(a) = -1.591 \qquad f(x_m) = -0.0549 \qquad f(b) = 1.543$$
$$|b - a| = 0.25$$

4th iteration

$$a = 1.375 \qquad\qquad x_m = 1.4375 \qquad\qquad b = 1.5$$
$$f(a) = -0.0549 \qquad f(x_m) = 0.737 \qquad f(b) = 1.543$$
$$|b - a| = 0.125$$

5th iteration

$$a = 1.375 \qquad\qquad x_m = 1.406\,25 \qquad\qquad b = 1.4375$$
$$f(a) = -0.0549 \qquad f(x_m) = 0.339 \qquad f(b) = 0.737$$
$$|b - a| = 0.0625$$

6th iteration

$$a = 1.375 \qquad\qquad x_m = 1.390\,63 \qquad\qquad b = 1.406\,25$$
$$f(a) = -0.0549 \qquad f(x_m) = 0.142 \qquad f(b) = 0.339$$
$$|b - a| = 0.031\,25$$

7th iteration

$$a = 1.375 \qquad\qquad x_m = 1.382\,81 \qquad\qquad b = 1.390\,63$$
$$f(a) = -0.0549 \qquad f(x_m) = 0.0434 \qquad f(b) = 0.142$$
$$|b - a| = 0.015\,625$$

8th iteration

$$a = 1.375 \qquad\qquad x_m = 1.378\,91 \qquad\qquad b = 1.382\,81$$
$$f(a) = -0.0549 \qquad f(x_m) = -0.005\,72 \qquad f(b) = 0.0434$$
$$|b - a| = 0.007\,812\,5$$

9th iteration

$$a = 1.378\,91 \qquad\qquad x_m = 1.380\,86 \qquad\qquad b = 1.382\,81$$
$$f(a) = -0.005\,72 \qquad f(x_m) = 0.0188 \qquad f(b) = 0.0434$$
$$|b - a| = 0.003\,906\,25$$

10th iteration

$$a = 1.378\,91 \qquad x_m = 1.379\,89 \qquad b = 1.380\,86$$
$$f(a) = -0.005\,72 \qquad f(x_m) = 0.006\,61 \qquad f(b) = 0.0188$$
$$|b - a| = 0.001\,953\,125$$

11th iteration

$$a = 1.378\,91 \qquad x_m = 1.379\,40 \qquad b = 1.379\,89$$
$$f(a) = -0.005\,72 \qquad f(x_m) = 0.000\,445 \qquad f(b) = 0.006\,61$$
$$|b - a| = 0.000\,976\,562\,5$$

Since $|b - a| < 0.001$ after the 11th iteration, the computation ends and the final answer is assumed to be $x = 1.379\,40$, because

$$|f(x_m)| < |f(a)| < |f(b)|$$

If we had continued the computation, we would have obtained a final solution $x = 1.379\,365$ after 17 iterations. This is the correct value of the root, to 7 significant figures. However, the value $x = 1.379\,40$ represents an error of less than 0.01%.

In the above example we can see that the method of bisection requires more iterations than the methods presented earlier (see Examples 5.5 and 5.8). The method of bisection, however, does *not* make use of a trial-and-error procedure which may or may not converge. Hence divergence is not a problem. This is an important advantage of the method of bisection as compared with the earlier techniques.

Another interesting characteristic of the method of bisection is the fact that we can determine *in advance* how many iterations are required to reduce the interval to less than a prescribed size. For the first iteration the interval will be the original interval, of size $I_1 = |b - a|$. Therefore the interval sizes for the second, third, ..., nth iterations will be

$$I_2 = \frac{I_1}{2} \qquad I_3 = \frac{I_2}{2} = \frac{I_1}{2^2} \qquad \cdots \qquad I_n = \frac{I_1}{2^{n-1}}$$

For stopping, I_n must be less than ϵ, and so

$$\frac{I_1}{2^{n-1}} < \epsilon \qquad \text{or} \qquad n > \frac{\log (I_1/\epsilon)}{\log 2} + 1$$

The smallest positive integer value for n that satisfies this last inequality will be the required number of iterations.

Example 5.10. For the conditions of Example 5.9,

$$\frac{\log (I_1/\epsilon)}{\log 2} + 1 = \frac{\log [(2-1)/0.001]}{\log 2} + 1 = \frac{3}{0.301\,030} + 1 = 10.9658$$

The smallest integer that exceeds this value is 11; hence $n = 11$, as observed in Example 5.9.

5.3 SOLUTION OF SIMULTANEOUS LINEAR EQUATIONS

A system of simultaneous, linear algebraic equations can be written in general form as

$$a_{11}x_1 + a_{12}x_2 + \cdots + a_{1n}x_n = b_1$$
$$a_{21}x_1 + a_{22}x_2 + \cdots + a_{2n}x_n = b_2$$
$$\cdots\cdots\cdots\cdots\cdots\cdots\cdots\cdots\cdots\cdots\cdots\cdots$$
$$a_{n1}x_1 + a_{n2}x_2 + \cdots + a_{nn}x_n = b_n$$

The constants $a_{11}, ..., a_{nn}$ and $b_1, b_2, ..., b_n$ are assumed to be known. Our objective is to solve for the values of $x_1, x_2, ..., x_n$ that satisfy the above equations.

Gauss-Jordan Elimination

This procedure is much simpler and more efficient than the more familiar methods based upon direct algebraic substitution or the use of determinants. The basic idea is first to write the given constants in the form of an array,

$$\begin{bmatrix} a_{11} & a_{12} & \ldots & a_{1n} & b_1 \\ a_{21} & a_{22} & \ldots & a_{2n} & b_2 \\ \ldots\ldots\ldots\ldots\ldots\ldots\ldots \\ a_{n1} & a_{n2} & \ldots & a_{nn} & b_n \end{bmatrix}$$

and then to transform this array, row by row, until all the a's become 0's except the elements on the *principal diagonal* (upper left to lower right), which become 1's. The last column, which originally contained the b's, will then contain the values of the unknown x's. In other words, the original array will be transformed into

$$\begin{bmatrix} 1 & 0 & \ldots & 0 & x_1 \\ 0 & 1 & \ldots & 0 & x_2 \\ \ldots\ldots\ldots\ldots\ldots \\ 0 & 0 & \ldots & 1 & x_n \end{bmatrix}$$

Each transformation consists in adding some multiple of one row to another row, in such a manner that the desired result (a 0 or a 1) is obtained. The detailed procedure can best be illustrated by means of an example.

Example 5.11. Solve the following system of simultaneous linear equations using the Gauss-Jordan method:

$$3x_1 + 2x_2 - x_3 = 4$$
$$2x_1 - x_2 + x_3 = 3$$
$$x_1 + x_2 - 2x_3 = -3$$

We first express the given system of equations in the form of an array, i.e.

$$\begin{bmatrix} 3 & 2 & -1 & 4 \\ 2 & -1 & 1 & 3 \\ 1 & 1 & -2 & -3 \end{bmatrix}$$

The array is then transformed as follows.

1. Transform the *first column* to (1, 0, 0). To do so,

   (a) Multiply the first row by $-2/3$ and add to the second row. This results in a *new second row*:

   $$0 \qquad -7/3 \qquad 5/3 \qquad 1/3$$

   (Note that the multiplier, $-2/3$, is chosen so that a zero is obtained in the first column of the second row, i.e. $(-2/3)(3) + 2 = 0$.)

   (b) Multiply the first row by $-1/3$ and add to the third row. This results in a *new third row*:

   $$0 \qquad 1/3 \qquad -5/3 \qquad -13/3$$

   (Again, note that the multiplier, $-1/3$, is chosen so that a zero is obtained in the first column of the third row.)

   (c) Multiply the first row by $1/3$, resulting in a *new first row*:

   $$1 \qquad 2/3 \qquad -1/3 \qquad 4/3$$

   (The multiplier, $1/3$, is chosen so that a 1 is obtained in the first column of the first row.)

   As a result of these operations, our original array has become

   $$\begin{bmatrix} 1 & 2/3 & -1/3 & 4/3 \\ 0 & -7/3 & 5/3 & 1/3 \\ 0 & 1/3 & -5/3 & -13/3 \end{bmatrix}$$

2. Transform the *second column* to (0, 1, 0). To accomplish this,

   (*a*) Multiply the second row (of the new array) by 2/7 and add to the first row.   This results in a *new first row*:

$$1 \quad 0 \quad 1/7 \quad 10/7$$

   (*b*) Multiply the second row by 1/7 and add to the third row.   This results in a *new third row*:

$$0 \quad 0 \quad -10/7 \quad -30/7$$

   (*c*) Multiply the second row by $-3/7$, resulting in a *new second row*:

$$0 \quad 1 \quad -5/7 \quad -1/7$$

Thus, our array has become

$$\begin{bmatrix} 1 & 0 & 1/7 & 10/7 \\ 0 & 1 & -5/7 & -1/7 \\ 0 & 0 & -10/7 & -30/7 \end{bmatrix}$$

3. Transform the *third column* to (0, 0, 1). To do so,

   (*a*) Multiply the third row (of the new array) by 1/10 and add to the first row.   We now have a *new first row*:

$$1 \quad 0 \quad 0 \quad 1$$

   (*b*) Multiply the third row by $-1/2$ and add to the second row.   This results in a *new second row*:

$$0 \quad 1 \quad 0 \quad 2$$

   (*c*) Multiply the third row by $-7/10$, resulting in a *new third row*:

$$0 \quad 0 \quad 1 \quad 3$$

Our array has now become

$$\begin{bmatrix} 1 & 0 & 0 & 1 \\ 0 & 1 & 0 & 2 \\ 0 & 0 & 1 & 3 \end{bmatrix}$$

4. We can now read the solution directly from the last column.   Thus, $x_1 = 1$, $x_2 = 2$, $x_3 = 3$.

   It is easy to verify that (1, 2, 3) are the correct answers by simply substituting these values into the original three equations.

Gauss Elimination

A variant of the Gauss-Jordan technique is *Gauss elimination*, in which the original array

$$\begin{bmatrix} a_{11} & a_{12} & \ldots & a_{1n} & b_1 \\ a_{21} & a_{22} & \ldots & a_{2n} & b_2 \\ \multicolumn{5}{c}{\dotfill} \\ a_{n1} & a_{n2} & \ldots & a_{nn} & b_n \end{bmatrix}$$

is transformed into

$$\begin{bmatrix} c_{11} & c_{12} & \ldots & c_{1n} & d_1 \\ 0 & c_{22} & \ldots & c_{2n} & d_2 \\ \multicolumn{5}{c}{\dotfill} \\ 0 & 0 & \ldots & c_{nn} & d_n \end{bmatrix}$$

wherein all elements below the principal diagonal are 0's.   The x's are then obtained as

$$x_n = d_n/c_{nn}$$
$$x_{n-1} = (d_{n-1} - c_{n-1,n}x_n)/c_{n-1,n-1}$$
$$x_{n-2} = (d_{n-2} - c_{n-2,n-1}x_{n-1} - c_{n-2,n}x_n)/c_{n-2,n-2}$$
$$\dotfill$$
$$x_1 = (d_1 - c_{12}x_2 - c_{13}x_3 - \cdots - c_{1n}x_n)/c_{11}$$

In the successive transformations each of the 0's is obtained by adding some multiple of one row to another, as in the Gauss-Jordan method.   The details are illustrated in the example below.

Example 5.12.   Use Gauss elimination to solve the following system of equations (see Example 5.11):

$$3x_1 + 2x_2 - x_3 = 4$$
$$2x_1 - x_2 + x_3 = 3$$
$$x_1 + x_2 - 2x_3 = -3$$

Writing the equations in the form of an array, we obtain

$$\begin{bmatrix} 3 & 2 & -1 & 4 \\ 2 & -1 & 1 & 3 \\ 1 & 1 & -2 & -3 \end{bmatrix}$$

The array is now transformed as follows.

1.   Transform the *first column* to (3, 0, 0).   To do so,

   (a)   Multiply the first row by $-2/3$ and add to the second row.   This results in a *new second row*:

$$0 \quad -7/3 \quad 5/3 \quad 1/3$$

   (b)   Multiply the first row by $-1/3$ and add to the third row.   This results in a *new third row*:

$$0 \quad 1/3 \quad -5/3 \quad -13/3$$

Our original array has now become

$$\begin{bmatrix} 3 & 2 & -1 & 4 \\ 0 & -7/3 & 5/3 & 1/3 \\ 0 & 1/3 & -5/3 & -13/3 \end{bmatrix}$$

2.   Transform the *second column* to (2, $-7/3$, 0).   To do so, multiply the second row by $1/7$ and add to the third row, resulting in a *new third row*:

$$0 \quad 0 \quad -10/7 \quad -30/7$$

Our array can now be written

$$\begin{bmatrix} 3 & 2 & -1 & 4 \\ 0 & -7/3 & 5/3 & 1/3 \\ 0 & 0 & -10/7 & -30/7 \end{bmatrix}$$

Thus

$$c_{11} = 3 \quad c_{12} = 2 \quad c_{13} = -1 \quad d_1 = 4$$
$$c_{22} = -7/3 \quad c_{23} = 5/3 \quad d_2 = 1/3$$
$$c_{33} = -10/7 \quad d_3 = -30/7$$

3.   We can now solve for x_3, x_2, and x_1.

$$x_3 = d_3/c_{33} = (-30/7)/(-10/7) = 3$$
$$x_2 = (d_2 - c_{23}x_3)/c_{22} = [1/3 - (5/3)(3)]/(-7/3) = (-14/3)/(-7/3) = 2$$
$$x_1 = (d_1 - c_{12}x_2 - c_{13}x_3)/c_{11} = [4 - (2)(2) - (-1)(3)]/3 = 3/3 = 1$$

This is, of course, the same solution as obtained in Example 5.11, where we solved the same system of equations using the Gauss-Jordan method.

Although Gauss elimination may appear to be more complicated than the Gauss-Jordan method, it is actually simpler, since fewer calculations are required.   Also, numerical errors can be controlled more easily with this method.   Therefore Gauss elimination is usually preferred over Gauss-Jordan elimination, especially when solving large systems of equations on a computer.

Removing Zeros from the Principal Diagonal

Both the Gauss elimination and the Gauss-Jordan elimination methods will break down if a zero is encountered on the principal diagonal during the course of the computation.   Such zeros can easily be removed, however, by an appropriate interchange of rows.   (This interchange will have no effect on the final solution, since it is equivalent to a mere reordering of the equations.)   The computational procedure can then be continued in the usual manner.   (Note that *columns cannot be interchanged*, as this will alter the given system of equations.)   The procedure is illustrated in the following example.

Example 5.13.   Solve the following system of equations using Gauss elimination:

$$x_1 + 2x_2 - 2x_3 = -3$$
$$2x_1 + 4x_2 + x_3 = 4$$
$$x_1 - x_2 = 4$$

The initial array is

$$\begin{bmatrix} 1 & 2 & -2 & -3 \\ 2 & 4 & 1 & 4 \\ 1 & -1 & 0 & 4 \end{bmatrix}$$

We now transform this array in the following manner.

1.   Transform the *first column* to (1, 0, 0).   To do so,

(*a*)   Multiply the first row by −2 and add to the second row.   This results in a *new second row*:

$$0 \quad 0 \quad 5 \quad 10$$

(*b*)   Multiply the first row by −1 and add to the third row.   We obtain a *new third row*:

$$0 \quad -3 \quad 2 \quad 7$$

Our original array has now become

$$\begin{bmatrix} 1 & 2 & -2 & -3 \\ 0 & 0 & 5 & 10 \\ 0 & -3 & 2 & 7 \end{bmatrix}$$

The second row contains a zero on the principal diagonal.   We can correct this situation by interchanging the second and third rows, which results in

$$\begin{bmatrix} 1 & 2 & -2 & -3 \\ 0 & -3 & 2 & 7 \\ 0 & 0 & 5 & 10 \end{bmatrix}$$

(Notice that we still have (1, 0, 0) in the first column, as desired.)

2.   Ordinarily, we would transform the *second column* into the form $(c_{12}, c_{22}, 0)$.   In this case the transformation is not necessary, since the second column is already in the desired form.   Thus

$$c_{11} = 1 \qquad c_{12} = 2 \qquad c_{13} = -2 \qquad d_1 = -3$$
$$c_{22} = -3 \qquad c_{23} = 2 \qquad d_2 = 7$$
$$c_{33} = 5 \qquad d_3 = 10$$

3.   Solving for x_3, x_2, and x_1, we obtain

$$x_3 = d_3/c_{33} = 10/5 = 2$$
$$x_2 = (d_2 - c_{23}x_3)/c_{22} = [7 - (2)(2)]/(-3) = (3)/(-3) = -1$$
$$x_1 = (d_1 - c_{12}x_2 - c_{13}x_3)/c_{11} = [-3 - (2)(-1) - (-2)(2)]/1 = 3/1 = 3$$

Notice how easily we removed the zero from the principal diagonal in step 1, by interchanging the second and third rows.   Had we not carried out this interchange, we would not have been able to continue with steps 2 and 3.

Before leaving this section we again emphasize that the elimination procedures presented herein are far superior to the more common elementary methods, such as direct substitution or the use of determinants.   There are two reasons for this.   First, elimination procedures require fewer calculations than the elementary methods and are therefore more *efficient*.   This relative efficiency increases dramatically as the problem size increases; for large systems of equations, elimination techniques are enormously more efficient than the more elementary methods.   (Certain types of advanced engineering problems require the solution of *thousands* of simultaneous, linear algebraic equations.   Such problems are always solved with computers, often using Gauss or Gauss-Jordan elimination.)

Elimination methods are also more *accurate* than the elementary solution techniques, since the elementary methods often require the calculation of differences of nearly equal numbers.   Such calculations can be highly inaccurate.   Moreover, these inaccuracies can compound themselves, resulting in solutions that may be meaningless.   The elimination methods are much less prone to this type of difficulty.

5.4   GRAPHICAL INTEGRATION

Many technical problems require the evaluation of an integral.   Some of these problems can be solved using the standard integration techniques learned in calculus.   On the other hand, there are certain integration problems to which the customary methods do not apply.   In particular, standard integration techniques cannot be used if the integrand (i.e. the function to be integrated) is expressed graphically or numerically, rather than analytically.   Graphical integration can be very useful in such situations.

In order to evaluate the integral

$$I = \int_a^b f(x)\, dx$$

graphically, the integrand, $f(x)$, must be plotted against the independent variable, x, over the interval $a \leq x \leq b$.   The area under the curve is then carefully measured.   The desired value for I will equal this area.

One way to determine the area under an irregular curve is to approximate the actual area with a number of small, adjacent rectangles.   The area of each rectangle can easily be calculated (area = length × width), and the areas then added.   A reasonably accurate result (two or three significant figures) will be obtained if a large enough number of narrow rectangles is used.   This method is known as the *trapezoidal rule*.

Example 5.14.   Evaluate the integral

$$I = \int_0^1 x^2\, dx$$

using both standard and graphical integration techniques.   Compare the results obtained using the two methods.

This integral can, of course, be evaluated using the standard integration methods of calculus.

$$I = \int_0^1 x^2\, dx = \frac{1}{3} x^3 \Big|_0^1 = \frac{1}{3}(1-0) = \frac{1}{3}$$

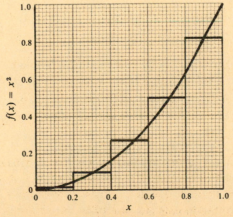

Fig. 5-7

Figure 5-7 shows a plot of $f(x)$ versus x, to be used in the graphical integration.   We see that the overall interval $0 \leqslant x \leqslant 1$ has been divided into 5 subintervals, each of width $\Delta x = 0.20$.   The height of each rectangle has been chosen in such a manner that the area *below the curve but above the rectangle* (on the right side of the subinterval) is roughly the same as the area *above the curve but below the rectangle* (on the left side of the subinterval).   In Table 5-1 the dimensions and the area of each rectangle are given.

Table 5-1

Subinterval No.	Width (Δx)	Height (h)	Area $(A = h \, \Delta x)$
1	0.20	0.02	0.004
2	0.20	0.10	0.020
3	0.20	0.27	0.054
4	0.20	0.50	0.100
5	0.20	0.82	0.164
		TOTAL	0.342

Adding up the areas, we obtain

$$I \approx A_1 + A_2 + A_3 + A_4 + A_5 = 0.004 + 0.020 + 0.054 + 0.100 + 0.164 = 0.342$$

This result differs from the true value, $I = 0.333$,   by less than 3%.

A more accurate answer could have been obtained if we had broken the overall interval into a larger number of subintervals (perhaps 10 or 20 subintervals).   However, the result obtained above may be sufficiently accurate for many engineering purposes.

The trapezoidal rule is not the only method that is available for determining the area under an irregular curve.   The area can also be measured directly using a *planimeter*, a special type of tracing instrument that is designed for this purpose.   Still another method for determining the area is to draw the curve carefully on heavy graph paper, and then cut out the area with a scissors and weigh it.   This weight is then compared with that of a unit square (or some multiple thereof).   For most applications, however, the trapezoidal rule will be adequate.   If greater accuracy is required, a suitable numerical procedure (as discussed in the next section) is usually selected rather than a more elaborate graphical integration technique.

5.5   NUMERICAL INTEGRATION

Numerical integration is very useful if an integrand is expressed in tabular form, or if the integrand is a function that cannot be integrated using the standard techniques learned in calculus.   In addition, numerical integration is sometimes used in place of graphical integration when a more accurate result is sought.

Trapezoidal Rule

The numerical version of the trapezoidal rule is simply an algebraic form of the graphical integration technique presented in Section 5.4.   If we wish to evaluate the integral

$$I = \int_a^b f(x) \, dx = \int_a^b y \, dx$$

by this method, we must have a set of n data points: $(x_1, y_1), (x_2, y_2), \ldots, (x_n, y_n)$, where $x_1 = a$ and $x_n = b$.   These data points will define $n - 1$ rectangular subintervals, where the width of the ith subinterval is given by

$$\Delta x_i = x_{i+1} - x_i$$

The height of the rectangle associated with a subinterval can be expressed as the arithmetic average of the ordinates at the endpoints, i.e.

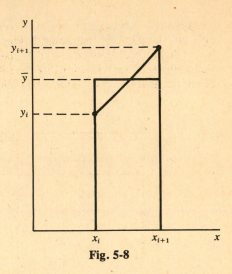

$$\bar{y}_i = \frac{y_i + y_{i+1}}{2}$$

(Note that the equal-areas condition will, in general, *not* be satisfied.) Hence the area of the rectangle is

$$A_i = \bar{y}_i \, \Delta x_i = \frac{(y_i + y_{i+1})(x_{i+1} - x_i)}{2}$$

These relationships are illustrated in Fig. 5-8. The given integral can now be expressed as

$$I = \int_a^b y \, dx \approx \sum_{i=1}^{n-1} A_i = \frac{1}{2} \sum_{i=1}^{n-1} (y_i + y_{i+1})(x_{i+1} - x_i)$$

Fig. 5-8

Example 5.15. The current passing through an electrical inductor can be determined from the equation

$$i = \frac{1}{L} \int_0^t v \, dt$$

where i = current, A
L = inductance, H
v = voltage, V
t = time, s

Suppose that a current of 2.15 A is induced over a period of 500 milliseconds. The variation of voltage with time during this period is given below. Determine the value of the inductance, using the above equation.

t, ms	0	5	10	20	30	40	50	60	70	80
v, V	0	12	19	30	38	45	49	50	49	47
t	90	100	120	140	160	180	230	280	380	500
v	45	42	36	33	30	27	21	16	9	4

In order to evaluate L in the above equation we must first evaluate the integral

$$I = \int_0^t v \, dt$$

If we use the trapezoidal rule we will have 19 intervals (since there are 20 data points), with the properties shown in Table 5-2. (Note that $\bar{v}_1 = (0 + 12)/2 = 6$ V, $\bar{v}_2 = (12 + 19)/2 = 15.5$ V, etc.) Thus,

$$I \approx \sum_{i=1}^{19} \bar{v}_i \, \Delta t_i = 10.7675 \text{ V·s}$$

We can now substitute this value into the given equation and solve for L. Thus,

$$L = \frac{1}{i} \int_0^t v \, dt \approx \frac{10.7675}{2.15} = 5 \text{ H}$$

If the values x_i are *equally spaced*, then the trapezoidal rule takes on a particularly simple form; namely,

$$I = \int_a^b y \, dx \approx \frac{\Delta x}{2} (y_1 + 2y_2 + 2y_3 + \cdots + 2y_{n-2} + 2y_{n-1} + y_n)$$

Table 5-2

Interval No.	Width, Δt (s)	Height, $\bar{v}$ (V)	Area, $\bar{v}\,\Delta t$ (V·s)
1	0.005	6.0	0.030
2	0.005	15.5	0.0775
3	0.01	24.5	0.245
4	0.01	34.0	0.340
5	0.01	41.5	0.415
6	0.01	47.0	0.470
7	0.01	49.5	0.495
8	0.01	49.5	0.495
9	0.01	48.0	0.480
10	0.01	46.0	0.460
11	0.01	43.5	0.435
12	0.02	39.0	0.780
13	0.02	34.5	0.690
14	0.02	31.5	0.630
15	0.02	28.5	0.570
16	0.05	24.0	1.200
17	0.05	18.5	0.925
18	0.10	12.5	1.250
19	0.12	6.5	0.780
		TOTAL	10.7675

or

$$I \approx \Delta x \left(\frac{y_1 + y_n}{2} + \sum_{i=2}^{n-1} y_i \right)$$

where Δx is the width of each subinterval. In this form the trapezoidal rule is especially easy to evaluate.

Example 5.16. Evaluate the integral

$$I = \int_0^1 e^{-x^2}\,dx$$

using the trapezoidal rule with 10 equal subintervals.

Since there are to be 10 subintervals, we will require $n = 11$ data points. Thus we construct Table 5-3.

Table 5-3

i	x_i	y_i
1	0	1.000
2	0.1	0.990
3	0.2	0.961
4	0.3	0.914
5	0.4	0.852
6	0.5	0.779
7	0.6	0.698
8	0.7	0.613
9	0.8	0.527
10	0.9	0.445
11	1.0	0.368

Each of the y-values after the first has been evaluated with an electronic calculator.   From the table,

$$\sum_{i=2}^{n-1} y_i = 6.779$$

Since $\Delta x = (1-0)/10 = 0.10$,   we can now evaluate the integral as

$$I = \int_0^1 e^{-x^2}\, dx \approx (0.10)\left(\frac{1.000 + 0.368}{2} + 6.779\right) = 0.7463$$

The correct value, to four significant figures, is 0.7468.   Thus the error is less than 0.1%.

It should be remarked that this particular integral *cannot* be evaluated by the standard methods of calculus.   The integral has, however, been evaluated numerically to a high degree of precision, and tabulated values are available in a number of reference books.

Simpson's Rule

Simpson's rule is a popular numerical integration technique which combines simplicity and accuracy. It is similar to the trapezoidal rule, in the sense that it approximates an irregular area with a number of sub-areas of known geometry.   Rather than erect rectangles on single intervals, however, we now pass a second-order polynomial (a parabola) through each successive group of three, adjacent, equally spaced data points, as indicated in Fig. 5-9.   The area under each polynomial can then be obtained by direct integration.

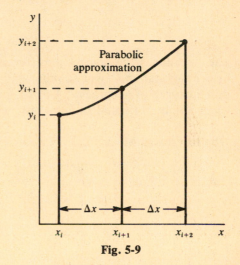

Fig. 5-9

If there is an even number of equal subintervals (i.e. an odd number of data points), then repeated use of this parabolic approximation results in the following simple expression:

$$I = \int_a^b y\, dx \approx \frac{\Delta x}{3}(y_1 + 4y_2 + 2y_3 + 4y_4 + 2y_5 + \cdots + 2y_{n-2} + 4y_{n-1} + y_n)$$

Notice the manner in which the interior y-values are alternately multiplied by 4 and 2.

Example 5.17.   Evaluate the integral

$$I = \int_0^1 e^{-x^2}\, dx$$

using Simpson's rule with 10 equal subintervals.

We have already evaluated this integral in Example 5.16, using the trapezoidal rule.   Therefore Table 5-4 can be constructed from the values given in Table 5-3.   Since there are 10 subintervals, we know that $\Delta x = 0.10$.   Therefore, the value of the integral can be approximated by

$$I \approx \frac{0.10}{3}(1.000 + 14.964 + 6.076 + 0.368) = 0.7469$$

The correct answer, to four significant figures, is 0.7468.

Recall that we obtained a value of 0.7463 in Example 5.16, using the trapezoidal rule.   Thus we see that Simpson's rule results in a more accurate answer, with very little additional computational effort.

When Simpson's rule is applicable it is usually preferred over the trapezoidal rule.   This is especially true when using an electronic calculator or a computer.   The reason is that the solutions obtained using Simpson's rule are more accurate than those obtained with the trapezoidal rule, but

Table 5-4

i	x_i	y_i	$4y_i$	$2y_i$
1	0	1.000		
2	0.1	0.990	3.960	
3	0.2	0.961		1.922
4	0.3	0.914	3.656	
5	0.4	0.852		1.704
6	0.5	0.779	3.116	
7	0.6	0.698		1.396
8	0.7	0.613	2.452	
9	0.8	0.527		1.054
10	0.9	0.445	1.780	
11	1.0	0.368		
		TOTALS	14.964	6.076

the difference in computational effort is practically negligible.   Equivalently, Simpson's rule requires less computation (fewer subintervals) than the trapezoidal rule in order to obtain answers with the same degree of accuracy.   On the other hand, Simpson's rule is restricted to problems having an even number of equal subintervals, whereas no such restriction exists for the trapezoidal rule.

The Integrated Mean

Many technical problems require the computation of an average for a given function, $f(x)$, that varies continuously over an interval $a \leq x \leq b$.   Such an average can be defined as

$$\overline{f} \int_a^b dx = \int_a^b f(x)\, dx$$

or

$$\overline{f} = \frac{\displaystyle\int_a^b f(x)\, dx}{\displaystyle\int_a^b dx} = \frac{1}{b - a} \int_a^b f(x)\, dx$$

The quantity $\overline{f}$ is known as the *integrated mean* of the function $f(x)$ over the interval $a \leq x \leq b$. It should be understood that the definition of the integrated mean has nothing to do with the manner in which the integral is evaluated.   Thus an integrated mean can be determined analytically (using the basic rules of calculus), graphically or numerically.

Example 5.18.   Determine an average value of the voltage for the voltage–time data presented in Example 5.15.

The average voltage is written

$$\overline{v} = \frac{1}{0.500 - 0} \int_0^t v\, dt$$

(Notice that the time is expressed in seconds rather than milliseconds, to be consistent with the units used in evaluating the integral in Example 5.15.)   We have already evaluated the integral numerically, and found the value to be $I \approx 10.7675$ V·s.   Therefore,

$$\overline{v} \approx \frac{10.7675}{0.500} = 21.535 \text{ V}$$

Solved Problems

5.1. Specify whether each of the following algebraic equations is linear or nonlinear.

$$(a) \quad 12x = 20 \qquad (b) \quad x^2 - 3x = 2 \qquad (c) \quad x + \cos x = 1 + \sin x$$

(a) linear, (b) nonlinear, (c) nonlinear.

5.2. Determine whether or not each of the equations presented in Problem 5.1 can be solved explicitly for x.

(a) $x = 5/3$.

(b) Using the quadratic formula

$$x = \frac{-b \pm \sqrt{b^2 - 4ac}}{2a}$$

we obtain

$$x = \frac{3 \pm \sqrt{9 + 8}}{2} = 3.562, -0.562$$

Hence $x_1 = 3.562$, $x_2 = -0.562$.

(c) Cannot be solved explicitly for x.

5.3. Another graphical procedure for solving a nonlinear algebraic equation is the following:

(1) Rearrange the equation into the form $x = g(x)$.

(2) Plot the two equations

$$y = x \quad \text{(a straight line)} \qquad y = g(x) \quad \text{(a curve)}$$

(3) Any value of x for which the curve and the straight line intersect will satisfy the equation $x = g(x)$, and hence be a root of the given equation.

Use this technique to solve

$$x^2 = 10 \cos x \qquad x > 0$$

(see Example 5.1).

We first rearrange the given equation into the form $x = \sqrt{10 \cos x}$. We then plot the equations

$$y = x \qquad y = \sqrt{10 \cos x}$$

as shown in Fig. 5-10. The two curves intersect at $x = 1.38$, which is the desired root. (This is the same value as obtained in Example 5.1.)

5.4. Rearrange the equation

$$x^5 - 3x^4 - 3x^2 - 35x + 60 = 0$$

into the form $x = g(x)$.

There are a number of different ways to rearrange the above equation into the desired form, e.g.

(a) $x = (x^5 - 3x^4 - 3x^2 + 60)/35$

(b) $x = (3x^4 + 3x^2 + 35x - 60)^{1/5}$

(c) $x = [(x^5 - 3x^4 - 35x + 60)/3]^{1/2}$

(d) $x = [(3x^2 + 35x - 60)/(x - 3)]^{1/4}$

(e) $x = 3 + 3x^{-2} + 35x^{-3} - 60x^{-4}$

(f) $x = [(35x - 60)/(x^3 - 3x^2 - 3)]^{1/2}$

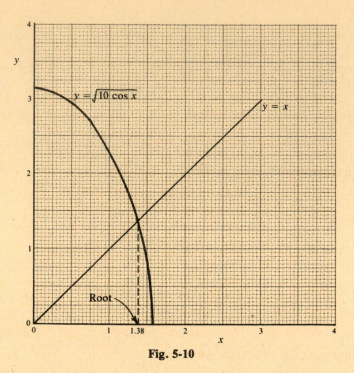

Fig. 5-10

5.5.   Show that the convergence criterion for the method of successive substitution is

$$\left| g'(\alpha) \right| < 1$$

where α represents the desired root of the equation $x = g(x)$.

   The difference between the ith approximation and the true solution (the ith *error term*) can be written as

$$e_i = x_i - \alpha = g(x_{i-1}) - g(\alpha)$$

If we expand $g(x_{i-1})$ in a *Taylor series* about α, we obtain

$$g(x_{i-1}) \approx g(\alpha) + g'(\alpha)(x_{i-1} - \alpha) = g(\alpha) + g'(\alpha)\, e_{i-1}$$

provided that x_{i-1} is sufficiently close to α.   Therefore

$$e_i = g(x_{i-1}) - g(\alpha) = g'(\alpha)\, e_{i-1}$$

From this last equation we can see that, if our starting value x_0 is not too far from α,

$$e_1 = g'(\alpha)\, e_0 \qquad e_2 = g'(\alpha)\, e_1 = [g'(\alpha)]^2\, e_0 \qquad \ldots \qquad e_n = [g'(\alpha)]^n\, e_0$$

In order for the method to converge, we require that

$$\lim_{n \to \infty} e_n = 0$$

This condition will be satisfied if and only if

$$\left| g'(\alpha) \right| < 1$$

5.6.   Show that the convergence criterion for the Newton-Raphson method is

$$\left| \frac{f(\alpha)\, f''(\alpha)}{[f'(\alpha)]^2} \right| < 1$$

where α represents the desired root of the equation $f(x) = 0$.

   Recall that Newton-Raphson method is a special case of the method of successive substitution, i.e.

$$x_i = g(x_{i-1}) = x_{i-1} - \frac{f(x_{i-1})}{f'(x_{i-1})}$$

We have already seen that the convergence criterion for the method of successive substitution is

$$|g'(\alpha)| < 1$$

Therefore the convergence criterion for the Newton-Raphson method must be

$$\left| \frac{d}{dx}\left[x - \frac{f(x)}{f'(x)} \right] \right|_{x=\alpha} < 1$$

Carrying out the differentiation, we obtain

$$\left| \frac{f(\alpha)\, f''(\alpha)}{[f'(\alpha)]^2} \right| < 1$$

5.7.   Solve the following system of equations using Gauss elimination.

$$
\begin{aligned}
0.1x_1 - 0.5x_2 \qquad\quad + \quad x_4 &= 2.7 \\
0.5x_1 - 2.5x_2 + \quad x_3 - 0.4x_4 &= -4.7 \\
x_1 + 0.2x_2 - 0.1x_3 + 0.4x_4 &= 3.6 \\
0.2x_1 + 0.4x_2 - 0.2x_3 \qquad\quad &= 1.2
\end{aligned}
$$

The given system of equations can be written as

$$
\begin{bmatrix}
0.1 & -0.5 & 0 & 1 & 2.7 \\
0.5 & -2.5 & 1 & -0.4 & -4.7 \\
1 & 0.2 & -0.1 & 0.4 & 3.6 \\
0.2 & 0.4 & -0.2 & 0 & 1.2
\end{bmatrix}
$$

We now perform the following operations:

1.   Transform the first column to $(0.1, 0, 0, 0)$ as follows:

   (a)   Multiply the first row by -5 and add to the second row.

   (b)   Multiply the first row by -10 and add to the third row.

   (c)   Multiply the first row by -2 and add to the fourth row.

   The result is

$$
\begin{bmatrix}
0.1 & -0.5 & 0 & 1 & 2.7 \\
0 & 0 & 1 & -5.4 & -18.2 \\
0 & 5.2 & -0.1 & -9.6 & -23.4 \\
0 & 1.4 & -0.2 & -2 & -4.2
\end{bmatrix}
$$

2.   Remove the zero from the principal diagonal by interchanging the second and third rows (or the second and fourth rows).

$$
\begin{bmatrix}
0.1 & -0.5 & 0 & 1 & 2.7 \\
0 & 5.2 & -0.1 & -9.6 & -23.4 \\
0 & 0 & 1 & -5.4 & -18.2 \\
0 & 1.4 & -0.2 & -2 & -4.2
\end{bmatrix}
$$

3.   Transform the second column to $(-0.5, 5.2, 0, 0)$ as follows: Multiply the second row by

$$-\frac{1.4}{5.2} = -0.269$$

and add to the last row.   The result is

$$
\begin{bmatrix}
0.1 & -0.5 & 0 & 1 & 2.7 \\
0 & 5.2 & -0.1 & -9.6 & -23.4 \\
0 & 0 & 1 & -5.4 & -18.2 \\
0 & 0 & -0.173 & 0.585 & 2.1
\end{bmatrix}
$$

4. Transform the third column to (0, −0.1, 1, 0) as follows: Multiply the third row by 0.173 and add to the fourth row.   The result is

$$\begin{bmatrix} 0.1 & -0.5 & 0 & 1 & 2.7 \\ 0 & 5.2 & -0.1 & -9.6 & -23.4 \\ 0 & 0 & 1 & -5.4 & -18.2 \\ 0 & 0 & 0 & -0.35 & -1.05 \end{bmatrix}$$

5. We now solve for x_4, x_3, x_2, and x_1 as follows.

$$x_4 = (-1.05)/(-0.35) = 3$$
$$x_3 = [-18.2 - (-5.4)(3)]/1 = -2$$
$$x_2 = [-23.4 - (-9.6)(3) - (-0.1)(-2)]/5.2 = 1$$
$$x_1 = [2.7 - (1)(3) - (0)(-2) - (-0.5)(1)]/0.1 = 2$$

5.8. Solve the system of equations presented in Problem 5.7 using Gauss-Jordan elimination.

The given system of equations can be written as

$$\begin{bmatrix} 0.1 & -0.5 & 0 & 1 & 2.7 \\ 0.5 & -2.5 & 1 & -0.4 & -4.7 \\ 1 & 0.2 & -0.1 & 0.4 & 3.6 \\ 0.2 & 0.4 & -0.2 & 0 & 1.2 \end{bmatrix}$$

We now perform the following operations:

1. Transform the first column to (1, 0, 0, 0) as follows:
   (a) Multiply the first row by −5 and add to the second row.
   (b) Multiply the first row by −10 and add to the third row.
   (c) Multiply the first row by −2 and add to the fourth row.
   (d) Multiply the first row by 10.

   The result is

$$\begin{bmatrix} 1 & -5 & 0 & 10 & 27 \\ 0 & 0 & 1 & -5.4 & -18.2 \\ 0 & 5.2 & -0.1 & -9.6 & -23.4 \\ 0 & 1.4 & -0.2 & -2 & -4.2 \end{bmatrix}$$

2. Remove the zero from the principal diagonal by interchanging the second and third rows (or the second and fourth rows).

$$\begin{bmatrix} 1 & -5 & 0 & 10 & 27 \\ 0 & 5.2 & -0.1 & -9.6 & -23.4 \\ 0 & 0 & 1 & -5.4 & -18.2 \\ 0 & 1.4 & -0.2 & -2 & -4.2 \end{bmatrix}$$

3. Transform the second column to (0, 1, 0, 0) as follows:
   (a) Multiply the second row by $5/5.2 = 0.962$ and add to the first row.
   (b) Multiply the second row by $-1.4/5.2 = -0.269$ and add to the last row.
   (c) Multiply the second row by $1/5.2 = 0.192$.

   The result is

$$\begin{bmatrix} 1 & 0 & -0.096 & 0.77 & 4.5 \\ 0 & 1 & -0.019 & -1.85 & -4.5 \\ 0 & 0 & 1 & -5.4 & -18.2 \\ 0 & 0 & -0.173 & 0.585 & 2.1 \end{bmatrix}$$

4. Transform the third column to (0, 0, 1, 0) as follows:

   (a)   Multiply the third row by 0.096 and add to the first row.

   (b)   Multiply the third row by 0.019 and add to the second row.

   (c)   Multiply the third row by 0.173 and add to the fourth row.

   The result is

   $$\begin{bmatrix} 1 & 0 & 0 & 0.25 & 2.75 \\ 0 & 1 & 0 & -1.95 & -4.85 \\ 0 & 0 & 1 & -5.4 & -18.2 \\ 0 & 0 & 0 & -0.35 & -1.05 \end{bmatrix}$$

5. Transform the fourth column to (0, 0, 0, 1) as follows:

   (a)   Multiply the fourth row by $0.25/0.35 = 0.714$ and add to the first row.

   (b)   Multiply the fourth row by $-1.95/0.35 = -5.571$ and add to the second row.

   (c)   Multiply the fourth row by $-5.4/0.35 = -15.429$ and add to the third row.

   (d)   Multiply the fourth row by $-1/0.35 = -2.86$.

   The result is

   $$\begin{bmatrix} 1 & 0 & 0 & 0 & 2 \\ 0 & 1 & 0 & 0 & 1 \\ 0 & 0 & 1 & 0 & -2 \\ 0 & 0 & 0 & 1 & 3 \end{bmatrix}$$

The desired answer now appears in the last column.   Thus, $x_1 = 2$, $x_2 = 1$, $x_3 = -2$ and $x_4 = 3$.

5.9.   When solving a system of simultaneous linear equations, it sometimes happens that the calculated solution is inaccurate because of numerical roundoff errors.   The following procedure can be used to improve the accuracy of the calculated solution.

1. Suppose that the calculated values of the unknown quantities are $\tilde{x}_1, \tilde{x}_2, ..., \tilde{x}_n$.   Substituting these values into the original system of equations, we obtain a corresponding set of values for the right-hand terms, i.e.

$$a_{11}\tilde{x}_1 + a_{12}\tilde{x}_2 + \cdots + a_{1n}\tilde{x}_n = \tilde{b}_1$$
$$a_{21}\tilde{x}_1 + a_{22}\tilde{x}_2 + \cdots + a_{2n}\tilde{x}_n = \tilde{b}_2$$
$$\cdots\cdots\cdots\cdots\cdots\cdots\cdots\cdots\cdots$$
$$a_{n1}\tilde{x}_1 + a_{n2}\tilde{x}_2 + \cdots + a_{nn}\tilde{x}_n = \tilde{b}_n$$

The calculated values $\tilde{b}_1, \tilde{b}_2, ..., \tilde{b}_n$ will differ from the original values $b_1, b_2, ..., b_n$ because of the inaccuracies in the $\tilde{x}$'s.

2. Calculate the quantities

$$d_1 = b_1 - \tilde{b}_1$$
$$d_2 = b_2 - \tilde{b}_2$$
$$\cdots\cdots\cdots\cdots$$
$$d_n = b_n - \tilde{b}_n$$

3. Now solve the system of equations

$$a_{11}e_1 + a_{12}e_2 + \cdots + a_{1n}e_n = d_1$$
$$a_{21}e_1 + a_{22}e_2 + \cdots + a_{2n}e_n = d_2$$
$$\cdots\cdots\cdots\cdots\cdots\cdots\cdots\cdots\cdots$$
$$a_{n1}e_1 + a_{n2}e_2 + \cdots + a_{nn}e_n = d_n$$

for the unknown quantities $e_1, e_2, ..., e_n$, using the same solution procedure that was used to calculate the $\tilde{x}$'s.   These e's will be *correction terms* for the $\tilde{x}$'s.

4. Once the e's have been determined, calculate a new (corrected) set of x's as follows.

$$x_1 = \tilde{x}_1 + e_1$$
$$x_2 = \tilde{x}_2 + e_2$$
$$\cdots\cdots\cdots\cdots$$
$$x_n = \tilde{x}_n + e_n$$

5. The entire procedure can be repeated if desired.   Usually, however, this will not be necessary.

Suppose that the following system of equations had been solved inaccurately:

$$3x_1 + 2x_2 - x_3 = 4$$
$$2x_1 - x_2 + x_3 = 3$$
$$x_1 + x_2 - 2x_3 = -3$$

to yield the approximate solution $x_1 = 1.2$, $x_2 = 1.8$, $x_3 = 2.7$. (The correct solution is $x_1 = 1$, $x_2 = 2$, $x_3 = 3$, as shown in Examples 5.11 and 5.12.) Use the above procedure to obtain a more accurate solution.

We have $\tilde{x}_1 = 1.2$, $\tilde{x}_2 = 1.8$, $\tilde{x}_3 = 2.7$. Thus we calculate the $\tilde{b}$'s as

$$\tilde{b}_1 = (3)(1.2) + (2)(1.8) - (1)(2.7) = 4.5$$
$$\tilde{b}_2 = (2)(1.2) - (1)(1.8) + (1)(2.7) = 3.3$$
$$\tilde{b}_3 = (1)(1.2) + (1)(1.8) - (2)(2.7) = -2.4$$

The d's can now be calculated as

$$d_1 = 4 - 4.5 = -0.5$$
$$d_2 = 3 - 3.3 = -0.3$$
$$d_3 = -3 - (-2.4) = -0.6$$

We must now solve the system of equations

$$3e_1 + 2e_2 - e_3 = -0.5$$
$$2e_1 - e_2 + e_3 = -0.3$$
$$e_1 + e_2 - 2e_3 = -0.6$$

The solution is $e_1 = -0.20$, $e_2 = 0.20$, $e_3 = 0.30$. Therefore, the corrected solution to the original system of equations is

$$x_1 = 1.2 + (-0.2) = 1.0$$
$$x_2 = 1.8 + 0.2 = 2.0$$
$$x_3 = 2.7 + 0.3 = 3.0$$

5.10.   Evaluate the integral

$$I = \int_0^1 x^2\, dx$$

graphically, using 10 rectangles of equal width. Compare the solution with the results obtained in Example 5.14.

Figure 5-11 shows a plot of x^2 versus x. We see that the overall interval $0 \le x \le 1$ has been divided into 10 subintervals, each of width $\Delta x = 0.10$. The height and area of each associated rectangle are given in Table 5-5. The individual areas sum to 0.337, which differs from the true value, 0.333, by only 1.2%. This result compares favorably with the value 0.342, obtained by graphical integration with 5 subintervals in Example 5.14. Thus, the larger the number of subintervals, the more accurate the final result, as expected.

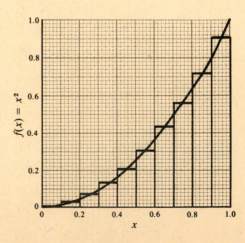

Fig. 5-11

Table 5-5

Subinterval No.	Height	Area
1	0.005	0.0005
2	0.025	0.0025
3	0.070	0.0070
4	0.130	0.0130
5	0.210	0.0210
6	0.310	0.0310
7	0.430	0.0430
8	0.560	0.0560
9	0.720	0.0720
10	0.910	0.0910
	TOTAL	0.3370

5.11. Evaluate the integral

$$I = \int_0^1 x^2 \, dx$$

numerically, using   (a) the trapezoidal rule, with 10 equal subintervals (i.e. $\Delta x = 0.10$);   (b) Simpson's rule, with $\Delta x = 0.10$.   Compare the results with the correct answer, $I = 1/3$.

(a) $I = (\Delta x)\left(\dfrac{y_1}{2} + y_2 + y_3 + \cdots + y_{10} + \dfrac{y_{11}}{2}\right)$

$\qquad = (0.10)\left[\dfrac{0}{2} + (0.1)^2 + (0.2)^2 + \cdots + (0.9)^2 + \dfrac{1}{2}\right]$

$\qquad = (0.10)(0 + 0.01 + 0.04 + 0.09 + 0.16 + 0.25 + 0.36 + 0.49 + 0.64 + 0.81 + 0.50)$

$\qquad = 0.335$

(b) $I = \dfrac{\Delta x}{3}(y_1 + 4y_2 + 2y_3 + 4y_4 + \cdots + 2y_9 + 4y_{10} + y_{11})$

$\qquad = \dfrac{0.10}{3}[0 + 4(0.01) + 2(0.04) + 4(0.09) + 2(0.16) + 4(0.25) + 2(0.36) + 4(0.49)$

$\qquad\qquad + 2(0.64) + 4(0.81) + 1]$

$\qquad = \dfrac{0.10}{3}(0 + 0.04 + 0.08 + 0.36 + 0.32 + 1.00 + 0.72 + 1.96 + 1.28 + 3.24 + 1.00)$

$\qquad = \dfrac{0.10}{3}(10.00) = \dfrac{1}{3}$

We see that the trapezoidal rule results in a value which is about $\frac{1}{2}\%$ too high (but is still very close to the exact answer).   Simpson's rule yields an answer which, in this case, is exact.   In fact, it can be shown that, whatever the value of Δx, Simpson's rule gives an exact answer if the integrand is a polynomial of degree 3 or less.   See Problem 5.48.

5.12. The following procedure can be used to improve the accuracy of a numerical integration when using the trapezoidal rule.

1. Evaluate the integral in the usual manner (using the trapezoidal rule), with n equal subintervals, each of width h.   Let us refer to the calculated value of the integral as I_h.

2. Repeat the procedure with m equal subintervals, each of width k.   Let the calculated value of the integral be referred to as I_k.

3. Assume that the *true* value of the integral can be expressed as

$$I = I_h + ch^2 \qquad \text{or} \qquad I = I_k + ck^2$$

where c is an unknown constant.   These two equations can be combined to eliminate c, resulting in

$$I = \frac{h^2 I_k - k^2 I_h}{h^2 - k^2}$$

This value will be closer to the true solution than either I_h or I_k.

Use the above procedure to improve the results of Example 5.14 and Problem 5.11(a) for

$$I = \int_0^1 x^2 \, dx$$

We have:

Example 5.14	Problem 5.11(a)
$n = 5$	$m = 10$
$h = 0.20$	$k = 0.10$
$I_h = 0.342$	$I_k = 0.335$

Substituting the results of the two numerical integrations into the correction equation, we obtain

$$I = \frac{(0.20)^2 (0.335) - (0.10)^2 (0.342)}{(0.20)^2 - (0.10)^2} = \frac{0.01340 - 0.00342}{0.03} = 0.333$$

This answer is exact, to three significant figures.

5.13. An automobile accelerates from a complete stop to some maximum speed in 10 s.   The velocity-versus-time curve can be represented by the equation

$$v = ct^{3/2}$$

where v = velocity, m/s
         c = constant of proportionality, $\text{m/s}^{5/2}$
         t = time, s

If the automobile travels 109.5 m during the 10 s interval, what is its final velocity?

The average (mean) velocity can be determined as

$$\bar{v} = \frac{109.5 \text{ m}}{10 \text{ s}} = 10.95 \text{ m/s}$$

But from the definition of the integrated mean,

$$\bar{v} = \frac{1}{10} \int_0^{10} v \, dt = \frac{1}{10} \int_0^{10} ct^{3/2} \, dt$$

$$= \left(\frac{2}{5}\right)\left(\frac{1}{10}\right) ct^{5/2} \Big|_0^{10} = \frac{c}{25} (10)^{5/2} = 12.65c$$

Therefore, $12.65c = 10.95$,   so that $c = 0.866 \text{ m/s}^{5/2}$.   We can now determine the final velocity as

$$v = 0.866(10)^{3/2} = 27.39 \text{ m/s}$$

Supplementary Problems

Answers to most problems are provided at the end of the book.

5.14.   Verify that $x = 1.38$ is an approximate root of $x^2 = 10 \cos x$ (see Example 5.1).

5.15.   Verify that $x = -2.2$, $x = 1.4$,   and $x = 3.6$ are approximate roots of the equation

$$x^5 - 3x^4 - 3x^2 - 35x + 60 = 0$$

(see Example 5.4).

5.16.   Solve graphically:

   (a) $x^5 + 3x^2 - 10 = 0$ (d) $4x^3 - x^2 + 3x - 10 = 0$
   (b) $x^3 + 2x - 7 = 0$ (e) $2x = \tan x$ (smallest positive root)
   (c) $x^2 - 4x + 4 = 0$ (f) $x + \cos x = 1 + \sin x$ $\pi/2 < x < \pi$

5.17.   Obtain all real roots of the equation

$$x^5 - 3x^4 - 3x^2 - 35x + 60 = 0$$

using the graphical procedure described in Problem 5.3 (see also Example 5.4).

5.18.   Verify that $x = 1.379\,365$ is an accurate value for the positive root of $x^2 = 10 \cos x$ (see Examples 5.5, 5.8 and 5.9).

5.19.   Solve each of the following equations using the method of successive substitution.   Rearrange each equation so that the computation will converge.   For equations (a) through (f), compare the solutions with those obtained in Problem 5.16.

   (a) $x^5 + 3x^2 - 10 = 0$ (e) $2x = \tan x$ (smallest positive root)
   (b) $x^3 + 2x - 7 = 0$ (f) $x + \cos x = 1 + \sin x$ $\pi/2 < x < \pi$
   (c) $x^2 - 4x + 4 = 0$ (g) $0.125x^3 + 0.925x + 0.586\,25 = 0$
   (d) $4x^3 - x^2 + 3x - 10 = 0$ (h) $-0.375x^3 + 5.50x^2 - 18.156\,25 = 0$

5.20.   Solve the equations given in Problem 5.19 by the Newton-Raphson method.   For each equation compare the rate of convergence using the two methods.

5.21.   Solve the equations given in Problem 5.19 by the method of bisection.   Compare the amount of computation required to solve each equation with the amounts required using successive substitution and the Newton-Raphson method.

5.22.   The equation $x^5 - 3x^4 - 3x^2 - 35x + 60 = 0$ has been shown to have three real roots, at approximately $x = -2.2$, $x = 1.4$,   and $x = 3.6$ (see Example 5.4).   If the equation is rearranged into the form

$$x = \frac{x^5 - 3x^4 - 3x^2 + 60}{35}$$

show that the method of successive substitution can be used to solve for the root whose value is approximately 1.4, but cannot be used to obtain the other two roots.   (Do not actually solve.)

5.23. Show that the Newton-Raphson method can be used to obtain all three real roots of the equation

$$x^5 - 3x^4 - 3x^2 - 35x + 60 = 0$$

(Recall that the roots are approximately $x = -2.2$, $x = 1.4$, and $x = 3.6$.) Do not actually solve the equation.

5.24. Suppose that the equation $x^5 - 3x^4 - 3x^2 - 35x + 60 = 0$ is to be solved for the real root between $x = 0$ and $x = 3$, using the method of bisection. (a) How many iterations will be required to reduce the interval that contains the root to less than 0.001? (b) If 16 iterations are carried out, what will be the size of the final interval? What will be the accuracy of the final answer?

5.25. The voltage drop across an electronic device is given by

$$v = 2e^{-0.1t} \sin 0.5t$$

where v represents the voltage drop (millivolts) and t represents time (milliseconds). (a) Determine how long it will take for the voltage to reach a value of 0.5 mV. (b) Determine the maximum voltage drop and the time when this occurs. (c) Determine the minimum voltage drop (i.e. the "largest negative" voltage drop) and the time when this occurs.

5.26. The *specific volume* of water (m³/kg) varies with water temperature. Within the limits of 0 °C and 33 °C the specific volume can be determined by the formula

$$V = 0.9999 \times 10^{-3} - 6.43 \times 10^{-8}\, T + 8.51 \times 10^{-9}\, T^2 - 6.8 \times 10^{-11}\, T^3$$

Determine the temperature at which the specific volume will be a minimum. What will be the value of the specific volume at this temperature?

5.27. A gas storage tank has the shape of a cylinder with a hemispherical top. If the cylindrical portion of the tank has a height of 1.50 m, what must the radius be in order that the tank have a volume of 0.1 m³?

5.28. An electrical conductor is to be surrounded by a tubular insulator whose inner radius is r and outer radius is R. If the insulator is to be as efficient as possible, the ratio of the radii must satisfy the equation

$$\ln x = 1 - \frac{1}{x^2}$$

where $x = R/r$. (a) Determine the value of x that satisfies the above equation. (b) If the insulator has an inner radius of 5 mm, what will be its outer radius?

5.29. Home mortgage costs are determined by the equation

$$A = iP\, \frac{(1 + i)^n}{(1 + i)^n - 1}$$

where A = monthly payment (dollars)
           P = total amount of the loan (dollars)
           i = monthly interest rate expressed as a decimal (e.g. 6% per year would be $\frac{1}{2}$% per month, which would be expressed as 0.005)
           n = total number of monthly payments

(a) A person borrows $40 000 for 30 years. If the monthly payments are $307.57, what is the monthly interest rate? (b) What annual interest rate does this correspond to? (c) What will be the total cost of the loan (including interest) over the entire 30-year period?

5.30. Solve the following systems of simultaneous linear equations using Gauss elimination.

$$(a) \quad \begin{aligned} x_1 - 2x_2 + 3x_3 &= 17 \\ 3x_1 + x_2 - 2x_3 &= 0 \\ 2x_1 + 3x_2 + x_3 &= 7 \end{aligned}$$

$$(b) \quad \begin{aligned} 3x_1 + x_2 - 3x_3 &= -7 \\ x_1 \qquad + 2x_3 &= 0 \\ 2x_1 - 5x_2 - 2x_3 &= -16 \end{aligned}$$

$$(c) \quad \begin{aligned} 0.2x_1 + 1.5x_2 - 0.5x_3 \qquad &= -3.40 \\ 1.2x_1 \qquad + 1.5x_3 - 0.5x_4 &= 5.35 \\ x_1 - 0.2x_2 - 2.5x_3 + 1.8x_4 &= -9.05 \\ 0.8x_1 + 0.6x_2 + 2x_3 - x_4 &= 6.50 \end{aligned}$$

$$(d) \quad \begin{aligned} 11x_1 + 3x_2 \qquad + x_4 + 2x_5 &= 51 \\ 4x_2 + 2x_3 \qquad + x_5 &= 15 \\ 3x_1 + 2x_2 + 7x_3 + x_4 \qquad &= 15 \\ 4x_1 \qquad + 4x_3 + 10x_4 + x_5 &= 20 \\ 2x_1 + 5x_2 + x_3 + 3x_4 + 13x_5 &= 92 \end{aligned}$$

5.31. Solve each system of simultaneous linear equations given in Problem 5.30 using Gauss-Jordan elimination.

5.32. Solve the system of 5 simultaneous linear equations given in Problem 5.30(d) using the method of determinants (Cramer's rule). Compare the amount of effort required to obtain the solution with the amount of effort required when using Gauss elimination. Which is the better method?

5.33. Suppose that the following approximate solution has been obtained for the system of equations given in Problem 5.30(d): $x_1 = 3$, $x_2 = 2$, $x_3 = 1$, $x_4 = 0$, $x_5 = 6$. Use the method presented in Problem 5.9 to obtain a more accurate solution.

5.34. Use Gauss elimination to solve the least-squares equations in Problem 4.22 for c_1, c_2, and c_3.

5.35. A steel company manufactures four different types of steel alloys, called A1, A2, A3, and A4. Each alloy contains small amounts of chromium, molybdenum, titanium, and nickel. The required composition of each alloy is given below.

Alloy	Cr	Mo	Ti	Ni
A1	1.6%	0.7%	1.2%	0.3%
A2	0.6	0.3	1.0	0.8
A3	0.3	0.7	1.1	1.5
A4	1.4	0.9	0.7	2.2

Suppose that the ingredients are available in the following amounts:

Ingredient	Availability
Cr	1200 kg/day
Mo	800
Ti	1000
Ni	1500

What will be the production rate for each alloy, in metric tons per day (1 metric ton = 1000 kg)?

5.36. Figure 5-12 shows a complex network of electrical resistors. (a) Derive a set of equations for the current flows in the resistors, using Kirchhoff's laws. (b) Solve the equations for the current flows.

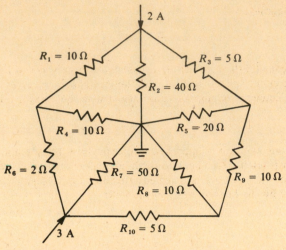

Fig. 5-12

5.37. A furnace wall is made up of three separate materials, as shown in Fig. 5-13. Let k_i represent the thermal conductivity of material i, and let Δx_i represent the corresponding thickness ($i = 1, 2, 3$). Then the steady-state equations of heat conduction can be written

$$T_0 - T_1 = (q\,\Delta x_1)/k_1$$
$$T_1 - T_2 = (q\,\Delta x_2)/k_2$$
$$T_2 - T_3 = (q\,\Delta x_3)/k_3$$
$$h_1(T_a - T_0) = h_2(T_3 - T_b)$$

where T_0, T_1, T_2, T_3, T_a, and T_b are temperatures, as illustrated in Fig. 5-13, q is the steady-state heat flux, and h_1 and h_2 are heat-transfer coefficients. Suppose that

$$\Delta x_1 = 0.5 \text{ cm} \qquad k_1 = 0.01 \text{ cal/cm} \cdot \text{s} \cdot {}^{\circ}\text{C}$$
$$\Delta x_2 = 0.3 \text{ cm} \qquad k_2 = 0.15 \text{ cal/cm} \cdot \text{s} \cdot {}^{\circ}\text{C}$$
$$\Delta x_3 = 0.2 \text{ cm} \qquad k_3 = 0.03 \text{ cal/cm} \cdot \text{s} \cdot {}^{\circ}\text{C}$$
$$T_a = 200 \text{ °C} \qquad h_1 = 1.0 \text{ cal/cm}^2 \cdot \text{s} \cdot {}^{\circ}\text{C}$$
$$T_b = 20 \text{ °C} \qquad h_2 = 0.8 \text{ cal/cm}^2 \cdot \text{s} \cdot {}^{\circ}\text{C}$$

and $q = 2.955 \text{ cal/cm}^2 \cdot \text{s}$. Solve for the unknown temperatures T_0, T_1, T_2, and T_3.

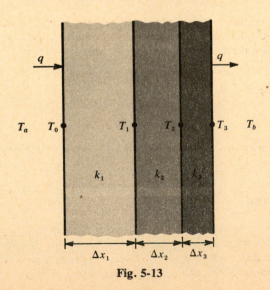

Fig. 5-13

5.38.   Evaluate the integral

$$I = \int_0^1 x^2 \, dx$$

by plotting x^2 against x ($0 \leqslant x \leqslant 1$) on a reasonably large sheet of heavy graph paper, cutting out the area beneath the curve and weighing it, and then weighing a piece of the same graph paper whose area is known (e.g. a square).   Compare your answer with the correct value, $I = 1/3$ (see also Example 5.14).

5.39.   A communications satellite is launched into orbit by a three-stage rocket.   During the launch the rocket's velocity is telemetered back to earth, giving the continuous record shown in Fig. 5-14.   Determine the distance the rocket has traveled after 60 s (a) by using the trapezoidal rule with 12 equal subintervals, (b) by the method of Problem 5.38.

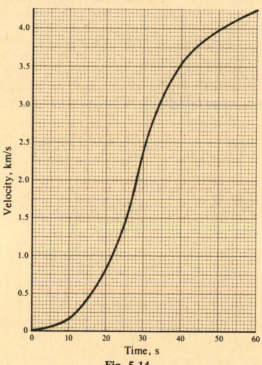

Fig. 5-14

5.40.   Plot the voltage-versus-time data presented in Example 5.15, and integrate graphically.   Determine the corresponding value for the inductance (L), and compare with the result obtained numerically in Example 5.15 using the trapezoidal rule ($L = 5$ H).

5.41.   Evaluate the integral

$$I = \int_0^1 e^{-x^2} \, dx$$

graphically.   Compare your answer with the results obtained using the trapezoidal rule and Simpson's rule (see Examples 5.16 and 5.17).

5.42.   Evaluate the integral

$$I = \int_0^1 e^{-x^2} \, dx$$

using the method discussed in Problem 5.38.   Compare your answer with the values obtained using the trapezoidal rule and Simpson's rule (see Examples 5.16 and 5.17).

5.43. Evaluate each of the following integrals graphically.

$$(a) \quad \int_1^2 \frac{dx}{x} \qquad (b) \quad \int_0^\pi x \sin x \, dx \qquad (c) \quad \int_{0.5}^{3.5} \frac{dx}{x^2 + 3} \qquad (d) \quad \int_{-2}^2 \frac{x \, dx}{5 + 2x}$$

5.44. Evaluate each of the integrals given in Problem 5.43 numerically, using the trapezoidal rule. Select a sufficiently small interval width so that each solution will be reasonably accurate.

5.45. Evaluate each of the integrals given in Problem 5.43 numerically, using Simpson's rule. Select a sufficiently small interval width so that each solution will be reasonably accurate. Compare with the results obtained using the trapezoidal rule, in Problem 5.44.

5.46. Evaluate the integral

$$I = \int_0^2 [1 - (x - 1)^2] \, dx$$

numerically, using the trapezoidal rule with (a) $\Delta x = 0.50$, (b) $\Delta x = 0.10$. Compare the results obtained with the true value, $I = 4/3$.

5.47. Obtain an improved value for the integral

$$I = \int_0^2 [1 - (x - 1)^2] \, dx$$

using the results obtained in Problem 5.46 and the method described in Problem 5.12.

5.48. Verify that Simpson's rule is exact for

$$I = \int_0^2 [1 - (x - 1)^2] \, dx$$

by choosing (a) $\Delta x = 0.50$, (b) $\Delta x = 0.10$. Compare the results with the true value, $I = 4/3$, and with the values obtained in Problem 5.46.

5.49. Obtain a mean value for y over the specified interval in each of the following cases.

(a) $y = x \sin x$ $0 \leqslant x \leqslant \pi$ (c) $y = x(5 + 2x)^{-1}$ $-2 \leqslant x \leqslant 2$
(b) $y = (x^2 + 3)^{-1}$ $0.5 \leqslant x \leqslant 3.5$ (d) $y = 1 - (x - 1)^2$ $0 \leqslant x \leqslant 2$

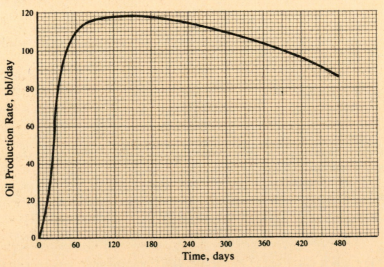

Fig. 5-15

5.50.   Figure 5-15 shows the production rate of an oil well as a function of time.   Determine   (*a*)   the total amount of oil produced during the first year (360 days) of operation,   (*b*)   the mean oil production rate during this period.

5.51.   Figure 5-16 shows a plot of an air pollutant against time for an entire day.   Determine the mean pollutant concentration   (*a*)   for the entire 24-hour period,   (*b*)   between 8:00 A.M. and 8:00 P.M.

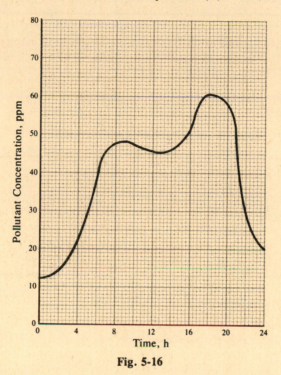

Fig. 5-16

5.52.   A rocket is fired vertically from the earth's surface.   During the firing the rocket's velocity and altitude are telemetered back to earth.   The data are plotted in Figure 5-17.

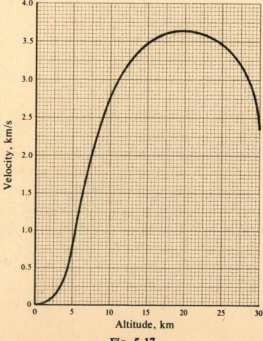

Fig. 5-17

Determine how long it takes for the rocket to reach an altitude of (*a*) 10 km, (*b*) 20 km, (*c*) 30 km. (*d*) What is the mean velocity during the time required for the rocket to reach an altitude of 30 km? (*e*) What is the mean velocity over the *distance* of 30 km?

5.53. A nuclear reactor contains fuel elements that are 10 mm thick. During steady-state operation, the temperature distribution within one of the fuel elements is as presented below.

x, mm	0	1	2	3	4	5	6	7	8	9	10
T, °C	172	406	616	784	891	928	891	784	616	406	172

Determine the mean temperature across the fuel element, by numerical integration.

5.54. The voltage in an electronic component varies with time, as shown below.

t, ms	0	0.1	0.2	0.3	0.4	0.6	0.8	1.0	1.5	2.0
v, mV	0	2.27	4.46	6.56	8.58	12.38	15.88	19.12	26.17	31.94
t	2.5	3.0	4.0	5.0	7.0	10.0	15.0	20.0	50.0	
v	36.66	40.53	46.29	50.15	54.47	56.94	57.86	57.98	58.00	

Determine the mean voltage during the 50 ms time period, using numerical integration.

5.55. The velocity of water flowing out the bottom of a cylindrical tank is given by the equation

$$v = 0.50 e^{-0.36 t}$$

where v = velocity, m/s
t = time, s

Suppose that the water flows through a circular hole whose radius is 1 cm. Determine, using numerical integration, the volume of water that leaves the tank during the first 30 s.

Chapter 6

Economic Calculations

A new engineering design frequently results in a potential investment opportunity for the sponsoring company. Thus the practicality of the design is often measured in economic rather than technical terms. Moreover, if several different designs have been proposed, resulting in a number of competing investment opportunities, the relative desirability of one investment over another is generally based upon certain economic criteria. The engineer must therefore have some knowledge of engineering economics in addition to a mastery of technical fundamentals.

6.1 SINGLE-PAYMENT INTEREST CALCULATIONS

Money is a commodity that is used to purchase goods and services. It therefore has *value*. If a person or an organization loans, deposits, or invests a sum of money, the lender is entitled to payment for the use of that money. Similarly, if a person or an organization borrows a sum of money, the borrower must pay for the use of the money. A sum of money that is loaned or borrowed is called *principal*; a payment that is made for the use of someone else's money is known as *interest*.

There are several different ways to compute interest, as discussed below. First, however, we must recognize that all interest calculations are based upon the use of an *interest rate*. The interest rate is ordinarily expressed as a percentage, but it must be converted to decimal form in order to be used in an interest calculation. To carry out this conversion, we first express any fractional percentage points in decimal form, and then divide the entire interest rate by 100.

Example 6.1. A bank pays $5\frac{1}{2}\%$ interest per year on its savings accounts. This interest rate must be expressed as 0.055 when it is used in an interest calculation.

Notice that we have not stated *how* the interest is paid (i.e. compounded annually, compounded quarterly, etc.). This will be discussed later in this section.

The determination of an appropriate interest rate is normally based upon two factors: the degree of risk associated with the intended investment, and general economic conditions. Risky investments require higher interest rates than conservative, safe investments.

Simple Interest

The amount of simple interest that accrues during each interest period (e.g. each year) is some fixed fraction of the principal. The total amount of interest is therefore equal to the number of interest periods times the interest accrued during each interest period. Thus,

$$I = niP$$

where
I = total amount of interest
n = number of interest periods
i = interest rate (expressed as a decimal)
P = principal

(Note that the amount of interest accrued during each interest period equals iP.)

Normally, when a simple-interest loan is made, nothing is repaid until the end of the loan period. At that time both the principal and the accumulated interest are repaid. The total amount due can be expressed as

$$F = P + I = P + niP = (1 + ni)P$$

Example 6.2. A student borrows $1000 from his uncle in order to finish school. His uncle agrees to charge him simple interest at the rate of $5\frac{1}{2}\%$ per year. If the student waits two years to repay the loan, how much will he have to repay?

Using the above formula,

$$F = [1 + 2(0.055)]($1000) = $1110$$

Thus the student must repay $1110 after two years. Of this amount, $1000 represents return of principal (the original amount of the loan), and $110 represents interest.

Example 6.3. A student has borrowed $1000 from a local bank in order to complete her education. She need not repay any of the loan for three years. At the end of three years, however, she must repay $1210, which is the entire cost of the loan. To what simple-interest rate does this correspond?

We know that $P = 1000, $F = 1210 and $n = 3$. We can therefore solve for i, resulting in

$$i = \frac{1}{n}\left(\frac{F}{P} - 1\right) = \frac{1}{3}\left(\frac{$1210}{$1000} - 1\right) = 0.07$$

Hence, the simple-interest rate is 7% per year.

Compound Interest

Technical economic calculations are usually based upon the use of *compound interest* rather than simple interest. When calculating compound interest, it is assumed that the interest accumulates from one interest period (e.g. one year) to the next.

During a given interest period, the current interest is determined as a percentage of the total amount owed (i.e. the principal plus the previously accumulated interest). Thus, for the first interest period, the interest is determined as

$$I_1 = iP$$

and the total amount due at the end of the first interest period is

$$F_1 = P + I_1 = P + iP = (1 + i)P$$

(as in simple interest). For the second interest period, the interest is determined as

$$I_2 = iF_1 = i(1 + i)P$$

and the total amount due at the end of the second interest period is

$$F_2 = P + I_1 + I_2 = P + iP + i(1 + i)P$$
$$= [1 + i + i(1 + i)]P = [(1 + i) + i(1 + i)]P = (1 + i)^2 P$$

For the third interest period,

$$I_3 = i(1 + i)^2 P \qquad F_3 = (1 + i)^3 P$$

and so on. In general, if there are n interest periods, it is easy to see that $F_n = (1 + i)^n P$. Usually the subscript is not included; we simply write

$$F = (1 + i)^n P$$

This is the well-known "law" of compound interest. Notice that F, the total amount due, increases *exponentially* with n, the number of interest periods.

Example 6.4. A student invests $1500 in a bank account paying 5% interest, compounded annually. If the student does not make any subsequent deposits or withdrawals, how much money will have accumulated after 20 years?

We know that $P = 1500, $i = 0.05$, and $n = 20$. Therefore

$$F = (1 + 0.05)^{20} ($1500) = $3979.95$$

The interest above is equal to $3979.95 - $1500 = $2479.95, which is substantially more than the original deposit.

It is interesting to see how much money would have accumulated after 20 years if simple interest had been paid. Then

$$F = [1 + (20)(0.05)]($1500) = $3000$$

which is significantly less than the amount accumulated through compound interest.

The ratio F/P is known as the *single-payment, compound-interest factor*. From the compound-interest formula,

$$F/P = (1 + i)^n$$

The factor $(1 + i)^n$ can easily be evaluated with an electronic calculator. This factor is usually presented in tabular form, however, as in Appendix B.

Example 6.5. A small manufacturing firm plans to borrow $500 000 for 4 years in order to expand its loading dock. If the company must pay interest at the rate of 7% per year, compounded annually, how much must the company repay at the end of the loan period?
    From Appendix B, we see that $F/P = 1.3108$ when $i = 7\%$ and $n = 4$.   Therefore, we obtain

$$F = (F/P)P = (1.3108)($500 000) = $655 400$$

We could, of course, have obtained the same result using $F = (1 + i)^n P$,   although the use of Appendix B is somewhat more convenient.

Example 6.6. A company wishes to borrow $75 000 to equip a product-testing lab.   If the company must repay $137 100 after 7 years, what annual interest rate is being charged?   Use Appendix B to obtain an answer.
    The single-payment, compound-interest factor can be determined as

$$F/P = $137 100/$75 000 = 1.8280$$

If we read across the row corresponding to $n = 7$ in Appendix B, we find that the factor 1.8280 appears in the column labeled 9%.   Therefore the interest rate is 9% per year, compounded annually.

Example 6.7. A company wishes to borrow $150 000 to promote a new product.   The current interest rate is 8% per year, compounded annually.   Suppose that the company does not plan to repay any of the loan until the end of the loan period, at which time the entire loan will be repaid.   If the company cannot repay more than $200 000, what is the maximum permissible loan period?
    The single-payment, compound-interest factor is obtained as

$$F/P = $200 000/$150 000 = 1.3333$$

If we read down the 8% column in Appendix B, we see that $F/P = 1.2597$ for $n = 3$,   and $F/P = 1.3605$ for $n = 4$.   Since the calculated value of F/P falls between these two values, we conclude that the loan period must be somewhat less than 4 years.
    A more accurate answer can be obtained using linear interpolation.   Thus,

$$n = 3 + \left(\frac{1.3333 - 1.2597}{1.3605 - 1.2597}\right)(4 - 3) = 3.73 \text{ years}$$

In practice, n is usually restricted to integer values.   Therefore the maximum permissible loan period would actually be 3 years.

6.2 EQUAL ANNUAL PAYMENTS

When money is accumulated in a savings account, the deposits are usually made at regular, periodic intervals (once a month, once a year, etc.).   Similarly, when money is borrowed, the loan is usually repaid in equal installments at regular intervals.   Therefore we must determine the manner in which a program involving equal periodic payments is affected by the interest rate.
    We begin by examining situations which involve equal *annual* payments.   Extensions to other time periods are considered in the problems at the end of this chapter.

Compound-Interest Factor

Consider a situation in which a given sum of money, say A, is deposited into a bank account at the end of each year. Suppose the money earns interest at a rate i, compounded annually. We wish to know how much money will have accumulated after n years.

To answer this question, we note that the first year's deposit will have increased in value to

$$F_1 = A(1 + i)^{n-1}$$

after n years. Similarly, the second year's deposit will have increased in value to

$$F_2 = A(1 + i)^{n-2}$$

and so on. The total amount of money that will have accumulated after n years can therefore be written as

$$
\begin{aligned}
F &= F_1 + F_2 + \cdots + F_n \\
&= A(1 + i)^{n-1} + A(1 + i)^{n-2} + \cdots + A \\
&= A[(1 + i)^{n-1} + (1 + i)^{n-2} + \cdots + 1]
\end{aligned}
$$

This equation can be expressed in a simpler form as

$$F = A\left[\frac{(1 + i)^n - 1}{i}\right]$$

(see Problem 6.2). Thus we can calculate the total accumulation (F) provided we know how much money is deposited each year (A), the interest rate (i), and the total number of years (n).

Example 6.8. Suppose that an engineer saves \$1000 each year for 30 years. At the end of each year the engineer deposits his savings in a bank account paying $5\frac{1}{2}\%$ interest, compounded annually. How much money will the engineer have accumulated after 30 years?

We are given that $A = \$1000$, $i = 0.055$, and $n = 30$. Therefore

$$F = (\$1000)\left[\frac{(1 + 0.055)^{30} - 1}{0.055}\right] = \$72\,435.48$$

Since the engineer has only deposited a total of \$30 000, we see that over half of the accumulated amount is the result of compounded interest.

The ratio F/A is known as the *equal-annual-payment, compound-interest factor*. This factor, like the single-payment, compound-interest factor, is often presented in tabular form as a function of i and n. Appendix C contains such tabular values of F/A. It is sometimes more convenient to use this table than to evaluate F/A with a calculator.

Example 6.9. A student plans to deposit \$600 into a savings account at the end of each year. If the bank pays 6% per year, compounded annually, how many years will be required for the student to accumulate \$10 000?

The ratio F/A can be computed as

$$F/A = \$10\,000/\$600 = 16.6667$$

If we examine the 6% column in Appendix C, we see that $F/A = 14.9716$ when $n = 11$, and $F/A = 16.8699$ when $n = 12$. Therefore, by linear interpolation,

$$n = 11 + \frac{16.6667 - 14.9716}{16.8699 - 14.9716}(12 - 11) = 11.89 \text{ years}$$

As a practical matter, however, the interest is not credited until the *end* of each year. Therefore 12 full years will be required for the student to fulfill his objective. After 12 years the student will have accumulated

$$F = (\$600)(16.8699) = \$10\,121.94$$

Capital-Recovery Factor

Let us now consider a somewhat different situation that involves equal annual payments. Suppose that a given sum of money, P, is deposited in a bank account where it earns interest at a rate

i, compounded annually. We wish to know the amount A that should be withdrawn at the end of each year so that the bank account will be depleted at the end of n years.

The initial deposit P could have increased in value to $P(1 + i)^n$ at the end of n years. On the other hand, the first withdrawal A could, if left in the bank, have increased to $A(1 + i)^{n-1}$ at the end of n years; the second withdrawal could have increased to $A(1 + i)^{n-2}$; ... ; the last withdrawal could have increased to $A(1 + i)^{n-n} = A$. Therefore

$$A(1 + i)^{n-1} + A(1 + i)^{n-2} + \cdots + A = P(1 + i)^n$$

$$A\left[\frac{(1 + i)^n - 1}{i}\right] = P(1 + i)^n$$

$$A = P\left[\frac{i(1 + i)^n}{(1 + i)^n - 1}\right]$$

Example 6.10. An engineer has accumulated $100\,000$ in a savings account paying $5\frac{1}{4}\%$ interest per year, compounded annually. Considering an early retirement, the engineer is interested in converting his $100\,000$ into a series of equal annual payments over a 20 year period. If he plans to withdraw the same amount of money each year, what is the maximum amount that he can withdraw?

Here we have $P = \$100\,000$, $i = 0.0525$, and $n = 20$. Therefore

$$A = (\$100\,000)\left[\frac{(0.0525)(1 + 0.0525)^{20}}{(1 + 0.0525)^{20} - 1}\right] = \$8195.23$$

Thus the engineer can withdraw $8195.23 at the end of each year, for 20 years. At the end of this time his entire savings account will be depleted. Notice, however, that he will have received a total of $163\,904.60 from his original $100\,000.

The factor A/P is called the *equal-annual-payment, capital-recovery factor*. Appendix D contains tabular values of A/P for various values of n and several common interest rates i.

Example 6.11. A person wishes to withdraw $1000 a year from a savings account paying 5% interest per year, compounded annually. If the withdrawals are to be made at the end of each year for 7 years, how much money must the person have in the bank at the start of the 7 year period?

To solve this problem let us write $P = A/(A/P)$. From Appendix D, $A/P = 0.1728$ when $n = 7$ and $i = 5\%$. Therefore,

$$P = \$1000/0.1728 = \$5787.04$$

Corporations invest money in business ventures, just as individuals deposit money in a savings account. It is assumed that a corporation is always able to earn a return on its money. Therefore, an equivalent interest rate is always associated with corporate investments.

Example 6.12. An electronics firm is planning to invest $1\,500\,000$ to develop a new type of computer terminal. It is assumed that the company will receive a return of $250\,000$ a year for ten years from the sale of these terminals, starting at the end of the first year. To what interest rate does this correspond, assuming annual compounding? Is this a good investment?

We can determine the capital-recovery factor as

$$A/P = \$250\,000/\$1\,500\,000 = 0.1667$$

From Appendix D, if we read across the row corresponding to $n = 10$, we see that $A/P = 0.1627$ for $i = 10\%$ and $A/P = 0.1770$ for $i = 12\%$. Therefore the interest rate corresponding to this planned investment is somewhat greater than 10% per year, compounded annually.

A more accurate result can be obtained by linear interpolation. Thus,

$$i = 10 + \left(\frac{0.1667 - 0.1627}{0.1770 - 0.1627}\right)(12 - 10) = 10.56\%$$

We cannot say, in absolute terms, that the proposed investment is either good or bad. This will depend on what alternatives may be available. If the company can invest its money in some other venture having a higher equivalent interest rate, then the current proposal does not appear attractive.

The equal-annual-payment, capital-recovery factor has another important interpretation, in conjunction with loans. Suppose that a person borrows a sum of money, P, at an interest rate i, compounded annually. If the person plans to repay the loan (including interest) with n equal annual payments, then the equal-annual-payment, capital-recovery factor is used to determine the amount of each payment. Many business loans (to corporations as well as individuals) are computed on this basis.

Example 6.13. A company plans to borrow \$150 000 at 9% per year, compounded annually, in order to finance the installation of air pollution equipment. The loan is to be repaid in 12 equal annual payments. Determine the amount of each payment.

This problem can be solved by writing $A = P(A/P)$. From Appendix D, $A/P = 0.1397$ for $n = 12$ and $i = 9\%$. Hence,

$$A = (\$150\,000)(0.1397) = \$20\,955$$

The company must therefore pay \$20 955 at the end of each year, for 12 consecutive years, in order to repay the entire loan. Note that the annual payments include interest as well as principal.

6.3 THE TIME VALUE OF MONEY

Because money has the ability to earn interest, its value increases with time. Thus, \$100 today is equivalent to \$105 one year from now if the interest rate is 5% per year, compounded annually. We therefore say that the *future worth* of \$100 is \$105 if $i = 5\%$, compounded annually, and $n = 1$.

Since money increases in value as we move from the present to the future, it must decrease in value as we move from the future to the present. Thus \$100 one year from now is equivalent to \$95.24 today if the interest rate is 5% per year, compounded annually. (Note that \$95.24 $\times$ 1.05 = \$100.) Thus, we say that the *present worth* of \$100 is \$95.24 if $i = 5\%$, compounded annually, and $n = 1$.

Example 6.14. A student will inherit \$5000 in three years. The student has a savings account that pays $5\frac{1}{2}\%$ per year, compounded annually. What is the present worth of the student's inheritance?

We can determine the present worth by writing $P = F/(F/P)$. Thus,

$$P = \$5000/(1 + 0.055)^3 = \$4258.07$$

In other words, if the student were to invest \$4258.07 in a savings account paying $5\frac{1}{2}\%$ per year, compounded annually, the student would have accumulated \$5000 at the end of three years.

We could, of course, have used Appendix B to solve this problem, although it would have been necessary to interpolate between the 5% and 6% entries.

Engineers are frequently required to calculate the present worth of a proposed investment in order to compare one investment strategy with another. We will say more about the comparison of investment strategies in the next section. First, however, it should be understood that the present-worth concept is not restricted to single-payment investments. The concept can be applied to equal annual payments, and to various combinations of single and equal annual payments. Cash inflows and cash outflows can both be included in a single proposed investment. Collectively, these various components are referred to as *cash flows*.

Example 6.15. An electronics firm is planning to manufacture a new type of car radio. An immediate investment of \$750 000 is required for manufacturing facilities, and an additional \$150 000 will be required for assembly-line modifications one year from now. Two years will be required to build adequate inventories. During the third year, the new radios are expected to generate a net income of \$180 000. This yearly income level is expected to continue for 10 years. At the end of this time (i.e. the end of the twelfth year) the product will be considered obsolete and production will cease. The original manufacturing facilities are then expected to have a salvage value of \$100 000. Determine the present worth of the proposed investment, based upon an equivalent interest rate of 8% per year, compounded annually.

We can think of this problem as consisting of four separate components: (1) the initial expenditure of \$750 000, (2) the secondary expenditure of \$150 000, (3) the ten equal annual inflows of \$180 000 each, and (4) the final inflow of \$100 000.   We will calculate the present worth of each item separately.

(1) The present worth of the initial expenditure is simply

$$P = -\$750\,000$$

where the minus sign indicates a cash outflow.

(2) The present worth of the secondary expenditure can be written as $P = F/(F/P)$.   From Appendix B, $F/P = 1.0800$ when $i = 8\%$ and $n = 1$.   Therefore

$$P = -\$150\,000/1.0800 = -\$138\,889$$

(3) The value of the 10 cash inflows *at the beginning of the 10 year period* (i.e. the start of the third year) can be written $P' = A/(A/P')$.   From Appendix D, $A/P' = 0.1490$ when $i = 8\%$ and $n = 10$.   Thus

$$P' = \$180\,000/0.1490 = \$1\,208\,054$$

Since this value corresponds to a time two years hence, we can obtain the present worth as

$$P = F/(F/P)$$

where $F = P'$.   From Appendix B, $F/P = 1.1664$ when $i = 8\%$ and $n = 2$.   Therefore we obtain

$$P = \$1\,208\,054/1.1664 = \$1\,035\,712$$

(4) The present worth of the salvage value is simply $P = F/(F/P)$.   From Appendix B, $F/P = 2.5182$ when $i = 8\%$ and $n = 12$.   Therefore

$$P = \$100\,000/2.5182 = \$39\,711$$

The present worth of the entire investment can now be obtained by summing the present worths of the separate components:

$$-\$750\,000 - \$138\,889 + \$1\,035\,712 + \$39\,711 = \$186\,534$$

When solving a problem which involves multiple cash inflows and outflows, it is helpful to chart the individual components as functions of time.   Cash outflows (expenditures) are written as negative quantities, whereas cash inflows are positive.   Time equivalences can be indicated by arrows.

Example 6.16.   The problem of Example 6.15 is charted in Fig. 6-1.   The same answer is obtained, but it is easier to visualize the problem when it is structured in this manner.

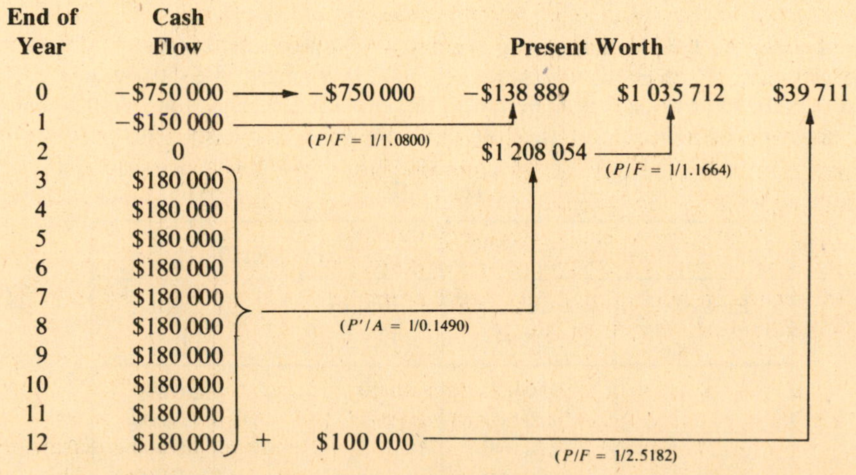

Fig. 6-1

A combination of cash inflows and outflows can be considered in terms of future worth as well as present worth, although the future-worth criterion is less common.

Example 6.17. Determine the future worth, at the end of the twelfth year, of the proposed investment described in Example 6.15. Figure 6-2 is the appropriate chart.

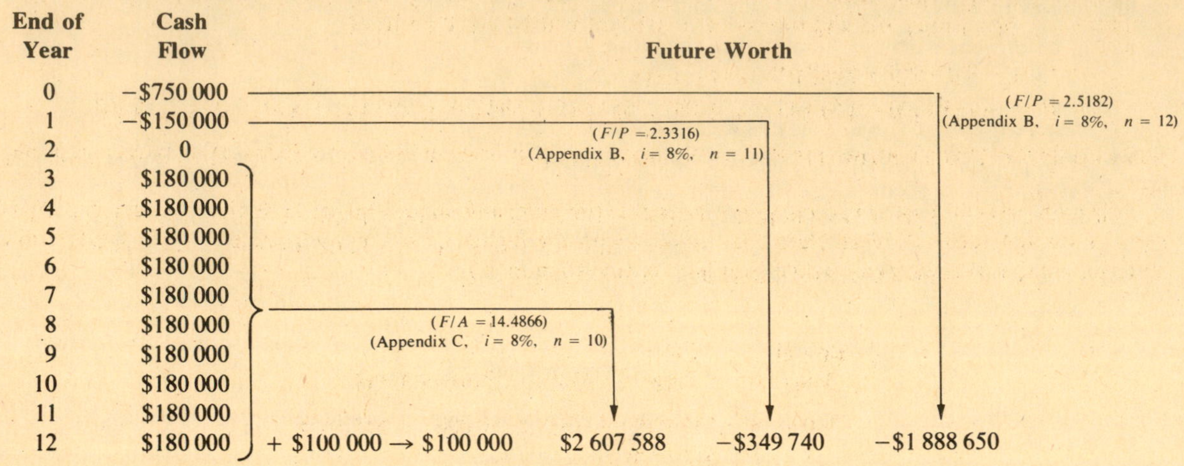

End of Year	Cash Flow	Future Worth			
0	−$750 000				$(F/P = 2.5182)$ (Appendix B, $i = 8\%$, $n = 12$)
1	−$150 000			$(F/P = 2.3316)$ (Appendix B, $i = 8\%$, $n = 11$)	
2	0				
3	$180 000				
4	$180 000				
5	$180 000				
6	$180 000				
7	$180 000				
8	$180 000	$(F/A = 14.4866)$ (Appendix C, $i = 8\%$, $n = 10$)			
9	$180 000				
10	$180 000				
11	$180 000				
12	$180 000	+ $100 000 → $100 000	$2 607 588	−$349 740	−$1 888 650

Fig. 6-2

The future worth of the proposed investment can now be obtained by summing the future worths of the separate components, as shown in the last line of the chart. Thus,

$$F = \$100\,000 + \$2\,607\,588 - \$349\,740 - \$1\,888\,650 = \$469\,198$$

We can also obtain the future worth of this investment by multiplying the present worth by an appropriate compound-interest factor. We have already determined that $P = \$186\,534$ (see Examples 6.15 and 6.16). Appendix B shows that $F/P = 2.5182$ when $i = 8\%$ and $n = 12$. Hence

$$F = (\$186\,534)(2.5182) = \$469\,730$$

which is essentially the same as the result obtained above. (The difference between the two values is due to rounding of the tabulated factors.)

6.4   COMPARING INVESTMENT STRATEGIES

Perhaps the most important idea in engineering economics is that of comparing one investment strategy with another, to determine which is more desirable. To do so, both investments must be expressed in terms of the same overall criterion, such as present worth. The investment having the higher present worth will be the more desirable.

Example 6.18. A company has developed two promising new products, but can afford to manufacture and market only one of them. Each product requires an initial investment of $350 000, and each is expected to provide a return of $720 000 over a 6 year period. The expected revenues are distributed differently with respect to time, however, as indicated by the cash flows shown in Table 6-1. If this company can normally

Table 6-1

Product A		Product B	
End of Year	Cash Flow	End of Year	Cash Flow
0	−$350 000	0	−$350 000
1	$120 000	1	$ 60 000
2	$120 000	2	$ 90 000
3	$120 000	3	$110 000
4	$120 000	4	$130 000
5	$120 000	5	$150 000
6	$120 000	6	$180 000

earn the equivalent of 10% per year, compounded annually, on its money, in which product should the company invest?

Let us determine the present worth of each proposed investment, using the chart procedure described in Section 6.3.   From Figs. 6-3 and 6-4, the two present worths are seen to be

$$P_A = -\$350\,000 + \$522\,648 = \$172\,648$$
$$P_B = -\$350\,000 + \$54\,545 + \$74\,380 + \$82\,645 + \$88\,792 + \$93\,138 + \$101\,603 = \$145\,103$$

Since product A yields a higher present worth than product B, the company should invest its money in product A.

(If both present worths had come out negative, the company should invest in neither product, but should earn its normal 10%.   Nevertheless, the product with the higher (less negative) present worth would still be superior, in that the company would lose less by investing in it.)

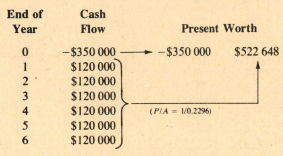

Fig. 6-3.   Product A

End of Year	Cash Flow	Present Worth					
0	−$350 000 → −$350 000	$54 545	$74 380	$82 645	$88 792	$93 138	$101 603
1	$ 60 000						
2	$ 90 000	($P/F = 1/1.1000$)					
3	$110 000	($P/F = 1/1.2100$)					
4	$130 000	($P/F = 1/1.3310$)					
5	$150 000	($P/F = 1/1.4641$)					
6	$180 000	($P/F = 1/1.6105$)					
		($P/F = 1/1.7716$)					

Fig. 6-4.   Product B

Future worth can be used instead of present worth when comparing one investment strategy with another.   However, future worth may be obtained simply by multiplying the present worth by a positive constant (i.e. F/P).   Therefore, if one proposal has a greater *present* worth than a competing proposal, then that proposal will also have a greater *future* worth than the competing proposal. In other words, the relative ranking of the two proposed investments will be the same, regardless of which decision criterion is selected.

Sometimes competing investments are projected over different time spans.   In such cases the economic criterion (e.g. the present worth) must be evaluated at the same point in time for each investment.

Example 6.19.   Determine which of the proposed investments in Table 6-2 is the more desirable, based upon the present-worth criterion.   Assume an interest rate of 12% per year, compounded annually.

In order to compare these two investments we will evaluate the present worth of each at the same point in time—namely, at the start of the first year (end of year 0).   See Figs. 6-5 and 6-6.

We find:

$$P_A = -\$500\,000 + \$607\,533 = \$107\,533$$
$$P_B = -\$714\,286 + \$765\,980 = \$51\,694$$

Table 6-2

Investment A		Investment B	
End of Year	Cash Flow	End of Year	Cash Flow
0	−$500 000	0	0
1	$200 000	1	−$800 000
2	$200 000	2	0
3	$200 000	3	$400 000
4	$200 000	4	$400 000
5	0	5	$400 000

End of Year	Cash Flow		Present Worth
0	−$500 000 ⟶ −$500 000		$607 533
1	$200 000		
2	$200 000		
3	$200 000	$(P/A = 1/0.3292)$	
4	$200 000		
5	0		

Fig. 6-5.   Investment A

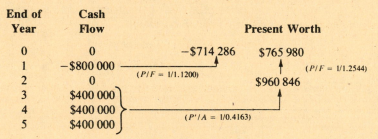

Fig. 6-6.   Investment B

The present worth of investment A is more than double that of investment B; hence investment A is more desirable than investment B.

It is interesting to observe that an entirely different conclusion is reached if we do not take into account the time value of money.   In fact, the net cash inflow of investment A (disregarding the 12% interest rate) is simply

$$(\text{cash inflow})_A = -\$500\,000 + 4(\$200\,000) = \$300\,000$$

and the net cash inflow of investment B is

$$(\text{cash inflow})_B = -\$800\,000 + 3(\$400\,000) = \$400\,000$$

Since B has a greater net cash inflow than A, we might be tempted to conclude that investment B is more desirable than investment A.   Such a conclusion would be incorrect.

Present worth and future worth are not the only commonly used economic decision criteria.   Other measures, such as *rate of return* and *return on investment*, are often used.   A discussion of these decision criteria is, however, beyond the scope of this book.

Solved Problems

6.1. Determine the value of the single-payment, compound-interest factor (F/P) for $n = 18$ and $i = 5\frac{1}{2}\%$, using (a) the formula for F/P, (b) Appendix B.

(a)
$$F/P = (1 + 0.055)^{18} = 2.6215$$

(b) The use of Appendix B is less straightforward, since there is no direct entry for $i = 5\frac{1}{2}\%$. We can, however, interpolate between the tabulated values for $i = 5\%$ and $i = 6\%$. Thus, using linear interpolation in the $n = 18$ row,

$$F/P = 2.4066 + \left(\frac{5.5 - 5.0}{6.0 - 5.0}\right)(2.8543 - 2.4066) = 2.6305$$

This value is close to, though not identical with, the value obtained in part (a).

6.2. Starting with the expression

$$F/A = (1 + i)^{n-1} + (1 + i)^{n-2} + \cdots + 1$$

for the equal-annual-payment, compound-interest factor, show that

$$F/A = \frac{(1 + i)^n - 1}{i}$$

Multiply the first of the above equations by $1 + i$. Thus

$$(F/A)(1 + i) = (1 + i)^n + (1 + i)^{n-1} + \cdots + (1 + i)$$

Now subtract the original equation from the above result.

$$(F/A)(1 + i) - (F/A) = [(1 + i)^n + (1 + i)^{n-1} + \cdots + (1 + i)]$$
$$- [(1 + i)^{n-1} + (1 + i)^{n-2} + \cdots + 1]$$

All the terms in brackets cancel, except the first and the last. Therefore, we have

$$(F/A)(i) = (1 + i)^n - 1 \qquad \text{or} \qquad F/A = \frac{(1 + i)^n - 1}{i}$$

6.3. (a) Determine an appropriate formula for the single-payment, compound-interest factor for the case where the interest is compounded quarterly rather than annually. Let n represent the number of years, and let i represent the annual interest rate, as before. (b) Generalize the result of (a) to the case of *continuous compounding*.

(a) The interest rate for each interest period will be $i/4$, and the number of interest periods will be $4n$. Therefore, the single-payment, compound-interest factor is

$$F/P = \left(1 + \frac{i}{4}\right)^{4n}$$

(b) For kn interest periods and an interest rate of i/k per period, we have

$$F/P = \left(1 + \frac{i}{k}\right)^{kn} = \left[\left(1 + \frac{i}{k}\right)^k\right]^n$$

Continuous compounding obtains in the limit as k, the number of interest periods per year, approaches ∞. But we know from calculus that as $k \to \infty$,

$$\left(1 + \frac{x}{k}\right)^k \to e^x$$

Hence

$$F/P \to [e^i]^n = e^{in}$$

Many financial institutions use continuous rather than discrete (i.e. annual, quarterly, etc.) compounding.

6.4. A student invests $1500 in a bank account paying 5% interest per year, compounded quarterly. If the student does not make any subsequent deposits or withdrawals, how much money will have accumulated after 20 years? Compare with the result obtained in Example 6.4, with annual compounding.

Using the formula developed in Problem 6.3(a), we have

$$F = (\$1500)\left(1 + \frac{0.05}{4}\right)^{(4)(20)} = (\$1500)(1 + 0.0125)^{80} = \$4052.23$$

If the interest is compounded annually rather than quarterly, only $3979.95 will have accumulated (see Example 6.4), which is $72.28 less than the result obtained above.

6.5. Determine formulas for the ratios F/A and A/P for the case where the interest is compounded monthly and the payments are made monthly rather than annually. Let n represent the number of years, and let i represent the annual interest rate.

The interest rate for each interest period will be $i/12$, and the number of interest periods will be $12n$. Therefore, the desired formulas are

$$F/A = \frac{[1 + (i/12)]^{12n} - 1}{i/12} \qquad \text{and} \qquad A/P = \frac{(i/12)[1 + (i/12)]^{12n}}{[1 + (i/12)]^{12n} - 1}$$

6.6. An engineer plans to borrow $50 000 in order to buy a house. Suppose the interest rate is 9% per year, compounded monthly. What will be the engineer's monthly payment if $n = 30$ years?

Using the formula for A/P presented in Problem 6.5, we have

$$A = (\$50\,000)\frac{(0.09/12)\,[1 + (0.09/12)]^{(12)(30)}}{[1 + (0.09/12)]^{(12)(30)} - 1} = (\$50\,000)\frac{(0.0075)(1 + 0.0075)^{360}}{(1 + 0.0075)^{360} - 1} = \$402.31$$

Thus the engineer will have to pay $402.31 each month, including interest.

6.7. Determine (*a*) the present worth, (*b*) the future worth, of the cash inflows and outflows listed below, based upon an interest rate of 10% per year, compounded annually.

End of Year	0	1	2	3	4	5
Cash Flow	−$25 000	$10 000	$12 000	$10 000	$15 000	$20 000

(*a*) See Fig. 6-7. The desired result is obtained by adding the individual components in the first row. Thus,

$$P = -\$25\,000 + \$9091 + \$9917 + \$7513 + \$10\,245 + \$12\,418 = \$24\,184$$

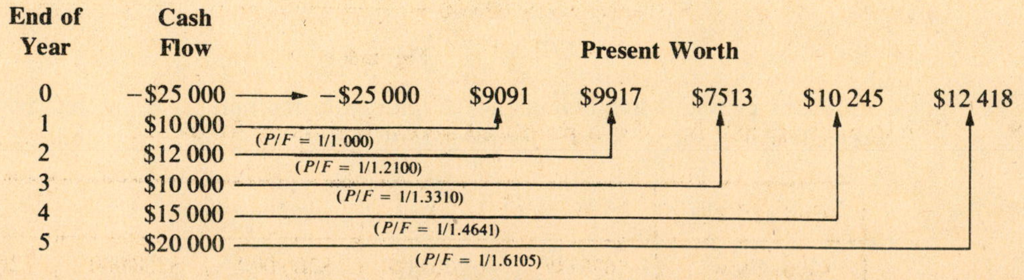

Fig. 6-7

(b) See Fig. 6-8. The future worth is obtained by adding the individual components in the last row. Hence,

$$F = \$20\,000 + \$16\,500 + \$12\,100 + \$15\,972 + \$14\,641 - \$40\,262 = \$38\,951$$

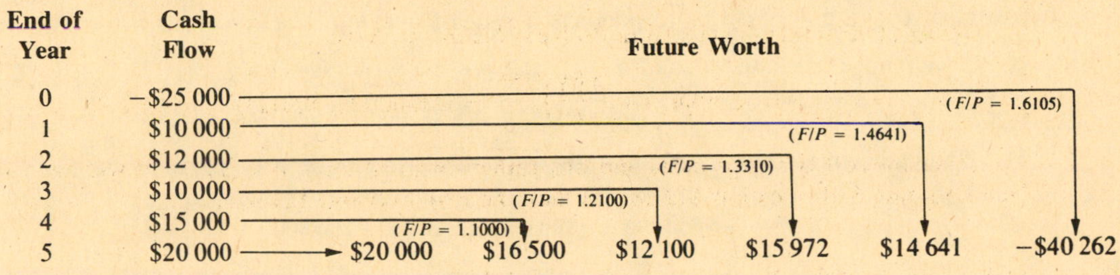

End of Year	Cash Flow	Future Worth

Fig. 6-8

We could also have obtained the future worth directly from part (a) by writing $F = P\,(F/P)$. Thus

$$F = (\$24\,184)(1.6105) = \$38\,948$$

which is essentially the same value as obtained above. (The compound-interest factors were obtained from Appendix B.)

6.8.   Let us once again examine the proposed investment described in Example 6.15. Suppose that the expected annual income is $120 000 rather than $180 000. What will be the present worth of the entire investment? What does the answer mean?

The problem can be charted as in Fig. 6-9. The present worth of the entire investment is obtained by adding the individual components in the first row. Thus,

$$P = -\$750\,000 - \$138\,889 + \$690\,474 + \$39\,711 = -\$158\,704$$

Notice that the final result is negative. This means that the company would be better off to invest its money in some other manner, based upon a return of 8% per year, compounded annually.

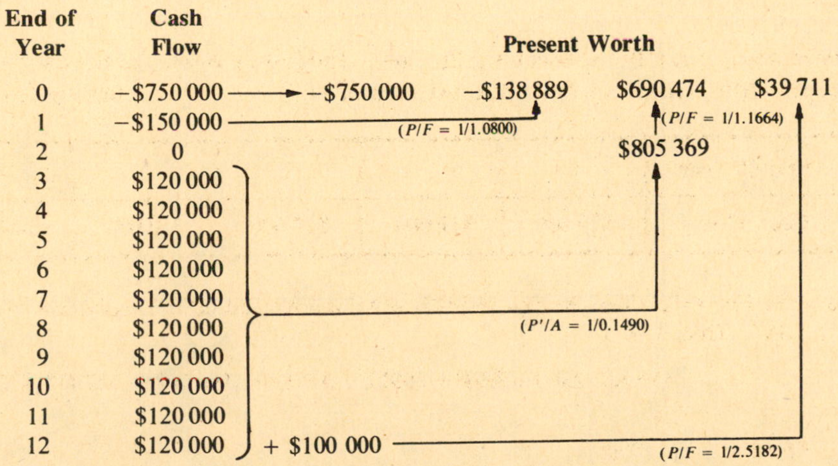

Fig. 6-9

6.9.   The expected cash flow for a proposed investment is as follows.

End of Year	0	1	2	3	4
Cash Flow	−$600 000	$200 000	$200 000	$200 000	$200 000

Determine the present worth of the proposed investment   (*a*)   ignoring the time value of money (i.e. $i = 0\%$);   (*b*)   assuming an equivalent interest rate of 5% per year, compounded annually;   (*c*)   assuming $i = 10\%$,   compounded annually;   (*d*)   assuming $i = 15\%$,   compounded annually;   (*e*)   assuming $i = 20\%$,   compounded annually.   Use the results of these calculations to prepare a plot of present worth versus interest rate.

(*a*)   If we ignore the time value of money, the present worth is simply

$$P = -\$600\,000 + 4(\$200\,000) = \$200\,000$$

(*b*)   When an interest rate is specified, the present worth can be obtained as

$$P = -\$600\,000 + (\$200\,000)(P/A) = -\$600\,000 + \frac{\$200\,000}{A/P}$$

The factor A/P can be obtained from the fourth row of Appendix D (since $n = 4$).   If $i = 5\%$, then $A/P = 0.2820$.   Hence

$$P = -\$600\,000 + \frac{\$200\,000}{0.2820} = \$109\,220$$

(*c*) $$P = -\$600\,000 + \frac{\$200\,000}{0.3155} = \$33\,914$$

(*d*) $$P = -\$600\,000 + \frac{\$200\,000}{0.3503} = -\$29\,061$$

(*e*) $$P = -\$600\,000 + \frac{\$200\,000}{0.3863} = -\$82\,268$$

Figure 6-10 shows a plot of present worth (P) versus interest rate (i).   Notice that the present worth decreases as the interest rate increases.   The curve crosses the abscissa (i.e. $P = 0$) when $i = 12.5\%$;   for values of i greater than 12.5%, the present worth is negative.   Such behavior is typical of investments of this type (i.e. an initial expenditure followed by several cash inflows).

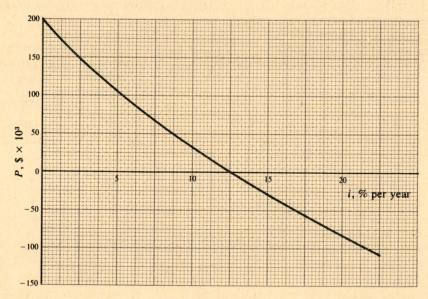

Fig. 6-10

Supplementary Problems

Answers to most problems are provided at the end of the book. There may be small discrepancies between answers obtained from Appendixes B, C, and D and answers obtained directly from the appropriate formulas.

6.10. Determine the factor F/P, based upon simple-interest calculations, for (*a*) $i = 5\%$, $n = 25$ years; (*b*) $i = 7\frac{1}{2}\%$, $n = 12$ years; (*c*) $i = 6\frac{3}{4}\%$, $n = 15$ years; (*d*) $i = 11\%$, $n = 6$ years; (*e*) $i = 12\frac{1}{2}\%$, $n = 20$ years; (*f*) $i = 5\frac{5}{8}\%$, n = 10 years.   Express your answers to 4-decimal accuracy.

6.11. Determine the factor F/P for each of the conditions listed in Problem 6.10, assuming that the interest is compounded annually. Compare each value with the corresponding value obtained in Problem 6.10. Are there significant differences in the results?

6.12. A student plans to borrow $2500 for 5 years at 8% per year, simple interest. No payments will be made during the 5 year period. At the end of 5 years, however, the student will repay the entire loan, including the interest that has accrued. (*a*) How much money will the student have to repay at the end of the 5 year period? (*b*) What is the equivalent compound-interest rate, based upon annual compounding?

6.13. Prepare a graph of F/P versus n based upon (*a*) simple interest, (*b*) compound interest (using Appendix B). Plot both equations on the same graph. Assume that $i = 10\%$ per year, and let n vary from 0 to 30 years. What conclusions can be reached about the nature of simple interest as compared with compound interest?

6.14. An engineer deposits $5000 in a savings account, where it will be allowed to collect interest for 20 years. No subsequent deposits or withdrawals will be made. Determine how much money will have accumulated at the end of 20 years, assuming an interest rate of (*a*) 5% per year, compounded annually; (*b*) 5% per year, compounded quarterly; (*c*) 6% per year, compounded annually; (*d*) 6% per year, compounded quarterly. How significant is the difference between annual and quarterly compounding? How significant is one additional percentage point in the interest rate?

6.15. A student borrows $3000 from a bank at $7\frac{1}{2}\%$ per year, compounded annually. The student will not make any payments for 4 years. At the end of 4 years, however, the student intends to repay the entire loan, including interest. Calculate the amount to be repaid, using (*a*) the single-payment, compound-interest equation, (*b*) Appendix B and interpolation.

6.16. In Problem 6.15, suppose that the interest is compounded monthly rather than annually. Use the appropriate form of the single-payment, compound-interest formula to solve for the amount of money owed after 4 years. Compare with the result obtained in Problem 6.15(*a*).

6.17. Prepare graphs showing how long it will take for a given sum of money to double, as a function of the interest rate, assuming the interest is compounded (*a*) annually, (*b*) quarterly, (*c*) monthly. Plot all three curves on the same graph (one curve for each type of compounding).

6.18. A group of engineers are forming a consulting company. They wish to borrow $20 000 for 3 years, in order to purchase office equipment and provide working capital. If the company must repay $25 194.24 at the end of 3 years, what interest rate is being charged? (Assume annual compounding.)

6.19. Suppose that a person deposits $5000 in a savings account that pays 6% interest. How much money will have accumulated after 25 years, assuming that the interest is compounded   (*a*)   annually, (*b*) quarterly,   (*c*) monthly,   (*d*) continuously [see Problem 6.3(*b*)]? What conclusions can be drawn about the frequency of compounding? Does continuous compounding offer a significant advantage over discrete compounding?

6.20. A company has recently developed a highly efficient, inexpensive, battery-powered lawn mower. At present the company's annual production rate is only 40% of total production capacity. However, the company expects that its annual sales (and hence its annual production rate) will increase by 9% each year, as public acceptance of battery-powered lawn mowers increases. If so, how long will it take for the company to be producing lawn mowers at its full capacity?

6.21. Determine the factor *F/A* under the following conditions, assuming annual compounding: (*a*) $i =$ 5%, $n = 25$ years; (*b*) $i = 7\frac{1}{2}\%$, $n = 12$ years; (*c*) $i = 6\frac{3}{4}\%$, $n = 15$ years; (*d*) $i = 11\%$, $n = 6$ years; (*e*) $i = 12\frac{1}{2}\%$, $n = 20$ years; (*f*) $i = 5\frac{5}{8}\%$, $n = 10$ years. Express your answers to 4-decimal accuracy.

6.22. An engineer deposits $1000 in a savings account at the end of each year. The bank pays 6% interest, compounded annually. (*a*)   How much money will have accumulated after 30 years? (*b*)   How long will it take the engineer to accumulate $100 000?

6.23. An engineer deposits a certain sum of money in a savings account at the end of each year. The bank pays 6% interest, compounded annually. How much money must the engineer save each year in order to accumulate $100 000 after 30 years?

6.24. An engineer deposits a certain sum of money in a savings account at the *beginning* of each year. The bank pays 6% interest, compounded annually. How much money must the engineer save each year in order to accumulate $100 000 after 30 years? Compare your answer with the result obtained in Problem 6.23.

6.25. A country has sufficient coal reserves to last for 400 years at the current annual rate of consumption. It is expected, however, that the rate of consumption will increase by 5% each year. If this is so, how long will it take for the country's coal reserves to be depleted?

6.26. Determine the factor *A/P* under the following conditions, assuming annual compounding: (*a*) $i =$ 5%, $n = 25$ years; (*b*) $i = 7\frac{1}{2}\%$, $n = 12$ years; (*c*) $i = 6\frac{3}{4}\%$, $n = 15$ years; (*d*) $i = 11\%$, $n = 6$ years; (*e*) $i = 12\frac{1}{2}\%$, $n = 20$ years; (*f*) $i = 5\frac{5}{8}\%$, $n = 10$ years. Express your answers to 4-decimal accuracy.

6.27. An engineer, who has accumulated $175 000 in a savings account, is planning an early retirement in order to teach part-time at a local university. The savings account earns interest at the rate of 6% per year, compounded annually. (*a*)   What is the largest fixed amount which the engineer can withdraw at the end of each year for 20 years? (*b*)   If the engineer withdraws $20 000 at the end of each year, how long will it take for the account to be depleted? (*c*)   If the engineer withdraws $10 500 at the end of each year, how long will it take for the account to be depleted?

6.28. A chemical company has developed a new lightweight, high-strength plastic. The company is planning to spend $6 000 000 for manufacturing facilities, in order to begin full-scale production. It is expected that the company will receive a return of $1 000 000 a year for 10 years. To what interest rate does this correspond, assuming annual compounding?

6.29.	A student plans to borrow $2500 at an interest rate of 8% per year, compounded annually. The loan will be repaid in 5 equal annual payments. (*a*) How much money will the student have to repay at the end of each year? (*b*) What will be the total amount of interest paid over the entire life of the loan?

6.30.	A philosopher has just won the $1 000 000 jackpot in the state lottery. He is given the choice of receiving $50 000 a year for 20 years, or accepting a lump-sum payment of $650 000. Suppose that he is able to deposit his money in a savings account that pays $5\frac{1}{2}\%$, per year, compounded annually. (*a*) What is the best course of action, if income taxes are neglected? (*b*) Is this decision likely to remain valid once the tax liabilities are considered?

6.31.	Most automobile loans are of the "add-on" type. With this kind of loan, the total cost of the loan is determined by the formula for simple interest. This total cost is then divided by the number of monthly payments, to obtain the amount due each month. Write an equation for the amount of each monthly payment.

6.32.	A recent engineering graduate is planning to purchase a luxury sports car. To do so she must borrow $6000, which she plans to repay in monthly installments over a period of 4 years. Two options are available:
(*a*)	She may obtain an "add-on" type automobile loan from a local bank, at an interest rate of 7% per year (see Problem 6.31).
(*b*)	She may obtain a conventional loan from her credit union, at an interest rate of 12% per year, compounded monthly (see Problem 6.5).
Determine which loan would be less expensive.

6.33.	An engineer plans to borrow $65 000 for 25 years in order to buy a house. If the interest rate is $8\frac{1}{2}\%$ per year, compounded monthly, find (*a*) the engineer's monthly payment (see Problem 6.6), (*b*) the total amount paid over the entire 25 year period.

6.34.	A furnace manufacturer has just developed a new energy-efficient furnace for commercial use. In order to begin production, the company plans to borrow $800 000 at an interest rate of 10% per year, compounded annually. The loan will be repaid in 7 equal annual payments. (*a*) Determine the amount of each annual payment. (*b*) What will be the total cost of the loan (i.e. the total amount of interest paid)? (*c*) What would be the total cost of the loan if the interest rate were 9% per year rather than 10%? Is the reduction of 1% significant?

6.35.	Determine the present worth of the following proposed investment.

End of Year	0	1	2	3	4	5	6
Cash Flow, $ $\times 10^3$	−1000	0	250	250	250	250	300

Select an annual interest rate of (*a*) 0%, (*b*) 4%, (*c*) 7%, (*d*) 10%. Assume annual compounding.

6.36.	Plot present worth versus annual interest rate using the results obtained in Problem 6.35. At what annual interest rate will the present worth be zero? What is the significance of a zero value for present worth?

6.37.	Calculate the future worth of the proposed investment given in Problem 6.35, for each of the specified interest rates. Obtain the future worth directly, rather than from the present worth. (The latter calculation can be used as a check, however.)

6.38. Plot future worth versus annual interest rate using the results obtained in Problem 6.37. At what annual interest rate will the future worth be zero? Compare with the zero value for present worth obtained in Problem 6.36.

6.39. Calculate the present and future worths of the following proposed investment, based upon an interest rate of 12% per year, compounded annually.

End of Year	0	1	2	3	4	5	6	7
Cash Flow, $ × 10^6	−2.5	−1.2	0.5	0.8	1.0	1.2	1.5	1.8

6.40. Determine which investment in Table 6-3 is the more desirable.

Table 6-3

Investment A		Investment B	
End of Year	Cash Flow	End of Year	Cash Flow
0	−$500 000	0	−$500 000
1	$150 000	1	$ 50 000
2	$150 000	2	$ 70 000
3	$150 000	3	$100 000
4	$150 000	4	$200 000
5	$150 000	5	$400 000

Base the calculations on annual compounding, using an interest rate of (*a*) 0% per year, (*b*) 5% per year, (*c*) 10% per year. Is the choice influenced by the interest rate?

6.41. Determine which investment in Table 6-4 is the more desirable, based upon an interest rate of 8% per year, compounded annually.

Table 6-4

Investment A		Investment B	
End of Year	Cash Flow	End of Year	Cash Flow
0	−$750 000	0	−$450 000
1	0	1	−$400 000
2	$150 000	2	$100 000
3	$150 000	3	$150 000
4	$200 000	4	$200 000
5	$200 000	5	$250 000
6	$200 000	6	$250 000
7	$200 000	7	$250 000

How would the two investments compare if the time value of money were not considered?

6.42. An automobile manufacturer plans to invest $27 000 000 in order to expand its production facilities. Two different proposals are being considered. The first proposal is to expand a current midwestern assembly plant. This will allow the manufacturer to produce an additional 50 000 automobiles per year. The manufacturer realizes an average net profit of $200 per automobile at this plant.

The second proposal is to build a new assembly plant on the west coast. This facility will have an annual output of only 40 000 automobiles per year, but the manufacturer expects to realize a profit of $300 per automobile.

The midwestern plant can begin production during the first year (i.e. the same year that the initial investment is made). However, the west coast plant requires a one-year start-up, so that production will not begin until the second year. Each assembly plant is assumed to have a lifetime of 5 years of actual production.

If the company uses an equivalent interest rate of 12% per year, compounded annually, which is the better proposal?

6.43. In Problem 6.42, suppose that the expected profit per automobile is reduced to $265 at the west coast plant because of higher labor and utility costs. Which proposal will now be more desirable?

6.44. A large regional office of a computer company requires an additional 400 000 square feet of storage space. One proposal is to erect a 400 000 ft² building now, at a cost of $12.50 per ft². Another proposal is to erect a 250 000 ft² building now, at a cost of $13.00 per ft², and then add 150 000 ft² 3 years from now, at a cost of $15.00 per ft². Which is the better proposal, based upon an annual interest rate of 8%?

6.45. For the situation described in Problem 6.44, suppose that the annual interest rate is 12% rather than 8%. Which would be the better proposal?

Chapter 7

Introduction to Computers

No single technological advancement has ever made as great an impact on the practice of engineering as the development of the digital computer. By using a computer, engineers are able to carry out an enormous number of calculations very quickly, easily, and inexpensively. This allows them to explore many different alternatives, thus refining their design calculations to the point where a superior product can be developed at a reasonable cost.

The earliest computers were very large, general-purpose devices. Computers of this type are still in common use; in fact, their costs have dropped markedly and their efficiency has increased enormously in recent years. Sophisticated scientific and engineering calculations are usually carried out on this type of computer.

In recent years there has been a trend toward smaller, less expensive computers (*minicomputers*). Many of these devices are small enough to fit on a desk top. Minicomputers are typically used for a variety of routine, less complicated, commercial and technical applications. Their use *complements* (rather than *replaces*) the use of the large, general-purpose computers.

Another very significant innovation has been the development of the *microcomputer*. A microcomputer is essentially a complete digital computer on a single electronic chip, no more than 4 mm by 4 mm square. These devices are widely used in process control and automated manufacturing. Many consumer items, such as digital watches, automatic cameras, and TV games, are microcomputer-based.

7.1 COMPUTER CHARACTERISTICS

Basically, a digital computer is an electronic device that can transmit, store, and manipulate information (*data*). Technical computations usually involve the processing of *numerical* data, although applications involving *character-type* data (e.g. names, addresses, etc.) are also quite common.

In order to process a particular set of data, the computer must be given an appropriate set of instructions (a *program*). These instructions are stored in a portion of the computer's *memory* for as long as they are needed.

At any time a stored program can be *executed*, causing the following things to happen.

1. A set of information, called the *input data*, will be entered into the computer (from a card reader, typewriter terminal, etc.) and stored in a portion of the computer's memory.

2. The input data will then be processed to produce certain desired results, known as the *output data*.

3. The output data (and perhaps some of the input data) will be printed onto a sheet of paper or displayed visually.

This 3-step procedure can be repeated many times if desired, thus causing a large quantity of data to be processed in rapid sequence.

Example 7.1. A computer has been programmed to calculate the area of a circle using the formula $A = \pi r^2$, given a numerical value for the radius, r, as input data. The steps involved would be as follows:

1. Read in the numerical value for the radius of the circle.
2. Calculate the value of the area, using the above formula.

3.   Print the values of the radius and the corresponding area.

4.   Stop.

Each of these steps might correspond to one instruction in a computer program.

We can represent the entire procedure pictorially, as shown in Fig. 7-1. This is known as a *flowchart*. Flowcharts can be very helpful in assisting the reader to visualize the flow of logic within a program.

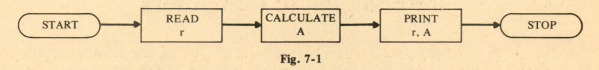

Fig. 7-1

The foregoing discussion illustrates two important characteristics of a digital computer: *memory* and *capability to be programmed*. A third important characteristic is its *speed and reliability*. We will say more about memory, speed, and reliability in the next few paragraphs. The subject of programmability will be discussed at length in the latter part of this chapter.

Memory

Modern computers have memories that range in size from a few thousand to several hundred thousand words, where a *word* may represent an instruction, a numerical quantity or a group of characters. (A *character* can be a letter, a digit, a comma, question mark, plus sign, etc.) The size of a computer's memory is usually expressed as some multiple of $2^{10} = 1024$ words. This quantity is referred to as 1K. Present-day microcomputers, such as the electronic chip shown in Fig. 7-2, generally have memories that range from 4K to 16K; 32K microcomputers are becoming increasingly common. Large scientific computers typically have memories of the order of 1000K.

Fig. 7-2

Speed and Reliability

Because of its extremely high speed, a computer can carry out calculations in just a few minutes that would require months—perhaps even years—if carried out by hand. Simple tasks, such as adding two numbers, can be carried out in a fraction of a microsecond (1 μs = 10^{-6} s). On a more practical level, the end-of-semester grades for all students in a large university can typically be processed in a few seconds of computer time, at a cost of about one dollar.

This very high speed is accompanied by an equally high level of reliability. Thus a computer practically never makes a mistake of its own accord. Highly publicized "computer" errors, such as a person's receiving a monthly bill of over one million dollars from a local department store, are almost always the result of a programming error or an error in data transmission.

7.2 MODES OF OPERATION

There are two different ways that a digital computer facility can be utilized. These are the *batch mode* and the *timesharing mode*. Both are very common. Each has its own advantages for certain types of problems.

Batch Processing

In *batch processing* a number of jobs are read into the computer and processed sequentially. (A *job* refers to a computer program and its associated data sets that are to be processed.) Usually the program and the data are recorded on punched cards; thus there will be a special deck of punched cards for each job. The information recorded on the cards will be read into the computer by means of a mechanical cardreader. The output, along with a listing of the computer program, will be printed on large sheets of paper by a high-speed printer.

Large quantities of information (both programs and data) can be transmitted into and out of the computer very quickly in batch processing. Therefore this mode of operation is well suited to jobs that require large amounts of computer time or are physically lengthy. On the other hand, the time required for a job to be processed in this manner may vary from several minutes to several hours, even though the job may have required only a second or two of actual computer time. (The job must wait its turn before it can be read, processed, and printed out.) Thus batch processing can be undesirable when it is necessary to process many small, simple jobs and return the results as quickly as possible.

Example 7.2. A student has 1000 different values for the radius of a circle and would like to calculate an area for each radius. To do so he must execute a computer program similar to the one described in Example 7.1 1000 times, once for each value of the radius (see Problem 7.1). Batch processing will be used, since all of the calculations are to be carried out in rapid succession. The program and the data will therefore be recorded on punched cards.

To process the data the student will read the deck of cards into the computer. The first part of the deck will contain the program, with one instruction per card. Following the program will be 1000 data cards, each of which will contain one value for the radius.

After the deck of cards has been read into the computer, there may be a delay of several hours before the program is executed and the input data processed. Once the computation has been completed, the results will be printed on a large sheet of paper. In this case the student will receive a listing that contains the actual program followed by 1000 pairs of numbers. Each pair of numbers will represent a value for the radius and its corresponding area. A portion of the computer-generated output is shown in Fig. 7-3.

Timesharing

Timesharing involves the simultaneous use of the computer by many different users. Each user is able to communicate with the computer through a *timesharing terminal* (or *typewriter terminal*), which may be connected to the computer by a telephone line or microwave circuit. These terminals are usually far from the computer—perhaps several hundred miles away. Because all the terminals can be served at essentially the same time, each user will be unaware of the others and will seem to have the entire computer at his own disposal.

```
R=  21.7873        A=  1491.27
R=  69.6209        A=  15227.5
R=  29.751         A=  2780.7
R=  96.3794        A=  29182.2
R=  46.3246        A=  6741.75
R=  76.7746        A=  18517.6
R=  82.9399        A=  21611.1
R=  18.1667        A=  1036.82
R=  15.9454        A=  798.773
R=  6.52568        A=  133.783
R=  79.3194        A=  19765.5
R=  64.4913        A=  13066.3
R=  92.7201        A=  27008.3
R=  89.4656        A=  25145.6
R=  97.4861        A=  29856.2
R=  80.4367        A=  20326.3
R=  99.2458        A=  30943.8
R=  68.785         A=  14864.
R=  61.9773        A=  12067.5
R=  73.1568        A=  16813.5
```

Fig. 7-3

A typical timesharing terminal is shown in Fig. 7-4. This type of terminal is sometimes called a "hard copy" terminal, since a permanent record of the timesharing session will be typed on paper. Other types of terminals use cathode-ray tubes (CRT's), similar to television screens, to display information. A permanent ("hard") record of the timesharing session cannot be obtained with a CRT-type terminal. Some newer terminals include both a CRT display and a capability for typing a permanent record.

A single timesharing terminal usually serves as both an input and an output device. The program and the input data are typed into the computer via the keyboard, and the output data are transmitted from the computer to the terminal where they are printed out or displayed. The transmission of data to and from the terminal is much slower, however, than the processing of the data by the computer. (The cardreaders and printers used in batch processing are able to transmit data much more rapidly than timesharing terminals.) It is this relative difference in speed that allows one computer to interact with several timesharing terminals simultaneously.

Timesharing is best suited for processing relatively simple jobs that do not require extensive data transmission or large amounts of computer time. Many of the computer applications that arise in schools and commercial offices have these characteristics. Such applications can be processed quickly, easily, and at minimum expense using timesharing.

Example 7.3. A large university has a computer timesharing capability consisting of 100 timesharing terminals located at various places around the campus. These terminals are connected to a large computer via telephone lines. Each terminal transmits data to or from the computer at a maximum speed of 30 characters per second. All the terminals can be used simultaneously, even though they are interacting with a single computer.

An additional 20 terminals are connected to the computer. Fifteen of these are located at 5 high schools in the area, and the remaining 5 terminals are stationed at a government research laboratory. All 120 terminals can be (and frequently are) used at the same time. By sharing the computer in this manner, each institution is able to utilize the services of a large computer at a reasonable cost.

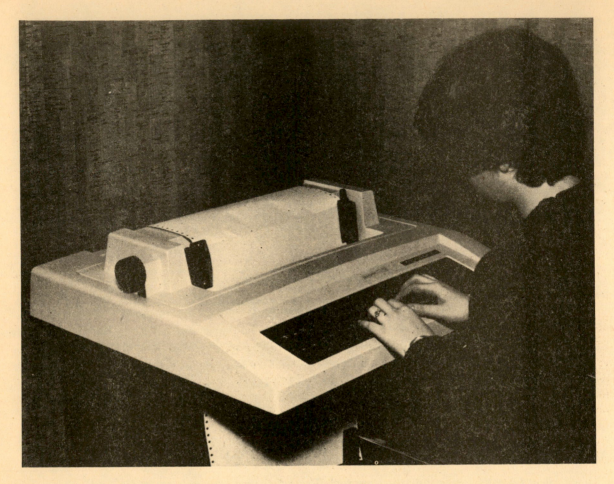

Fig. 7-4

Example 7.4.   A student has written a computer program which he wishes to execute in the timesharing mode.   To do so, he must type his program into a timesharing terminal, one instruction per line.   The instructions are transmitted to the computer, where they are stored in memory, as soon as they are typed.

After the entire program has been typed and stored in memory, the student can execute his program simply by typing the word RUN into the terminal.   The required input data must then be typed.   As soon as all of the input data have been entered, the output will automatically be computed and then transmitted directly to the terminal, where it will also be typed.

An important feature of timesharing is that the computer and the user may *interact* with each other during program execution.   Thus a relatively small portion of a program may be executed and a message printed out, indicating the preliminary results that have been computed and requesting additional data from the user.   Further execution of the program will be suspended until the additional data have been supplied.

In supplying the requested data, the user may be influenced by the preliminary results that have already been obtained.   Thus he may wish to study the output before supplying any additional data.   Moreover, the particular data supplied by the user may influence the manner in which the remainder of the program is executed.   Hence the computer and the user are, in a sense, conversing with each other, since the information supplied by one of the participants may affect the subsequent actions of the other.

Programs that make use of this idea are said to be written in a *conversational mode*.   Computerized games, such as tic-tac-toe, checkers, and chess, are excellent examples of interactive (or conversational mode) programs.

Certain features of the batch and timesharing modes can be combined if desired.   For example, it is possible to enter a set of input data directly from a timesharing terminal (thus eliminating the need for keypunching) and then proceed to process the data in the batch mode.   Another possibility is to use a cardreader (batch processing) to enter a program and a set of data, and then process the data in the timesharing mode.   Such "hybrid" operations are becoming increasingly common as computer systems grow in sophistication.

7.3 FUNDAMENTAL PROGRAMMING OPERATIONS

There are five fundamental programming operations that form the basis for virtually all computer programs, regardless of the particular problem being solved, the computer being used, or the programming language that has been selected.

Numerical Evaluation

Numerical evaluation involves the basic arithmetic operations of addition, subtraction, multiplication, division, and exponentiation.   The use of these five operations allows us to evaluate algebraic expressions numerically.

Logical Comparison

Logical comparison involves a comparison between two numerical quantities to see if they are equal, or to determine if one of the numbers is greater than the other.   (A logical comparison can also be made between two groups of characters to see whether or not they are equal, i.e. whether the first group is the same as the second group.)   Such comparisons are required in order to carry out branching operations and certain types of looping operations, as described below.

Example 7.5.   A computer program is being written which will calculate y as a function of x, where

$$y = \begin{cases} 0.5x & x \le 4 \\ 2x - 6 & x > 4 \end{cases}$$

Since one of two different equations will be selected, the program must contain a logical comparison between the given value for x and the constant 4.   Selection of the proper equation will depend upon the outcome of this comparison (see Example 7.6).

Branching

In a branching operation a transfer (a "jump") is made to one of several different places in a computer program, depending upon the outcome of a logical comparison.   This allows various "options" to be included within the program.

Example 7.6.   A flowchart for the program of Example 7.5 is shown in Fig. 7-5.   Notice that the logical comparison is enclosed within the diamond-shaped symbol, with a colon (:) separating the quantities that must be compared.   The program then branches to the left if $x \le 4$,   and to the right if $x > 4$.   The proper numerical evaluation is then carried out.

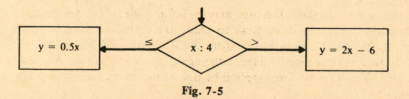

Fig. 7-5

Branching operations are included in most computer programs.   Many scientific and technical applications require several branching operations at various places within a single program.

Looping

In a looping operation some portion of a computer program is repeated many times in succession, until a specified stopping condition has been satisfied. The most common type of stopping condition involves the use of a *counter*, which increases in value by 1 after each pass through the loop. The looping operation then terminates once the counter has reached some specified value. A branching operation (i.e. a jump out of the loop) is used to bring about the actual termination.

Example 7.7. A portion of a computer program calculates the average of n different values for x, i.e.

$$x_{\text{avg}} = \frac{x_1 + x_2 + \cdots + x_n}{n}$$

assuming that n, the number of x-values, is known, and the actual values for x are stored in the computer's memory. The computation can easily be carried out if we utilize a loop to determine the cumulative sum

$$x_1 + x_2 + \cdots + x_n$$

Figure 7-6 shows a flowchart of the procedure. Initially, SUM is set equal to zero. Each pass through the loop then causes one of the x's to be added to SUM. This is indicated by writing SUM = SUM + x_i (i.e. the new value for SUM is equal to the old value plus the value of the current x_i). After n passes through the loop, SUM will represent the cumulative sum

$$x_1 + x_2 + \cdots + x_n$$

Notice that the counter, I, is initially assigned a value of 1, in preparation for the first pass through the loop. The counter then increases by 1 with each subsequent pass. This is represented as I = I + 1 (i.e. the new value for I is equal to the old value plus 1). The loop terminates after n passes (i.e. when I = n), as indicated by the logical comparison between I and n. The desired average is then obtained simply as SUM divided by n.

Another way to represent this looping operation is illustrated in Fig. 7-7. Notice that the actual loop is now enclosed within a dashed rectangle. The first block within the loop, which is labeled

FOR I = 1 TO n

indicates that the loop will be executed n times (I = 1 during the first pass, I = 2 during the second pass, etc., until I = n during the last pass). Thus it is not necessary to increment the counter (I = I + 1) or to include a branching operation (I : n) within the flowchart. It should be understood, however, that these operations are an integral part of the looping operation when it is shown in this manner.

This latter method for representing a loop (Fig. 7-7) is more desirable than the former method (Fig. 7-6), because it is simpler and more closely resembles the loop structures that are included in the commonly used

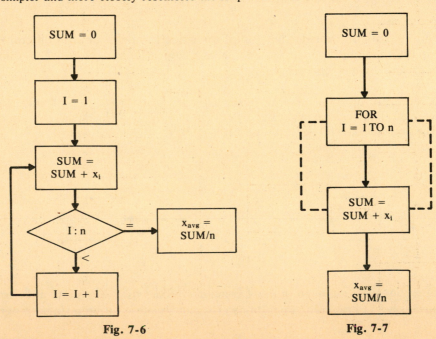

Fig. 7-6   Fig. 7-7

programming languages. We will use this latter representation for all finite loops (i.e. loops that are repeated a prescribed number of times) throughout the remainder of this chapter.

Another situation that arises frequently is the need to continue a looping operation until some particular condition has been satisfied (rather than continuing for a prescribed number of passes). This can be accomplished by utilizing a branching operation based upon some *calculated* quantity (rather than a counter) as a stopping condition.

Example 7.8. Suppose we wish to determine the least number of consecutive integers whose sum equals or exceeds some specified value. To do so, we will utilize a loop, as in Example 7.7. Now, however, the stopping condition takes the form $1 + 2 + 3 + \cdots + k \geq v$, where v represents the specified value. The desired number of integers is then K, the smallest k for which the stopping condition is met.

Figure 7-8 presents a flowchart of the procedure. We again see that SUM represents a cumulative sum, and I represents a counter, as in Example 7.7. Now, however, we add the consecutive values of the *counter* (the consecutive integers 1, 2, 3, ...) to SUM, and we compare each consecutive value of SUM with the specified value for v in order to establish a stopping condition. The final value of the counter will be the desired K.

Input/Output Operations

Input/output operations involve the transmission of data into the computer (from a cardreader, timesharing terminal, magnetic tape, etc.) or out of the computer (to a line printer, timesharing terminal, etc.). They are sometimes referred to as *read* and *write* operations. All computer programs require some form of input/output operations.

Example 7.9. In Example 7.7 we considered a portion of a computer program that calculates the average of n different values for x using the expression

$$x_{avg} = \frac{x_1 + x_2 + \cdots + x_n}{n}$$

Let us now complete the procedure by adding the required input/output statements.

Figure 7-9 presents a flowchart of the entire averaging procedure. Notice that the input operation (READ) is included within the loop, since a new value of x is required for each pass. On the other hand, only a single

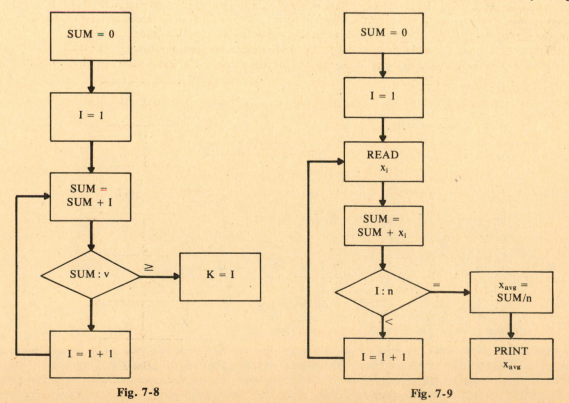

Fig. 7-8   Fig. 7-9

value (x_{avg}) must be printed out at the end of the computation.   Therefore the output operation (PRINT) appears at the end of the procedure, after the loop has been terminated.

7.4 DESIGNING A COMPUTER PROGRAM (FLOWCHARTING)

When writing a new computer program, the first step should always be the development of a complete program outline or flowchart.   This allows the analyst to specify exactly what features will be included in the program, and how the computation will be carried out, before becoming involved with detailed programming considerations.   Thus, the analyst is able to *design* the program *strategically*, independent of any particular programming language.

Example 7.10.   Design a computer program that will calculate the roots of the quadratic equation

$$ax^2 + bx + c = 0$$

where a, b, and c are known constants, such that a and b are not both zero.

The method used to carry out the computation will depend upon the assigned values for a, b, and c. Specifically:

1. If $a = 0$ (and $b \neq 0$), there will be one (real) root, given by $x = -c/b$.
2. If $a \neq 0$ and $b^2 = 4ac$,   there will be one (real, repeated) root, given by $x = -b/2a$.
3. If $a \neq 0$ and $b^2 > 4ac$,   there will be two real roots, given by

$$x_1 = \frac{-b + \sqrt{b^2 - 4ac}}{2a} \quad \text{and} \quad x_2 = \frac{-b - \sqrt{b^2 - 4ac}}{2a}$$

4. If $a \neq 0$ and $b^2 < 4ac$,   there will be two complex roots, given by

$$x_1 = \frac{-b + i\sqrt{4ac - b^2}}{2a} \quad \text{and} \quad x_2 = \frac{-b - i\sqrt{4ac - b^2}}{2a}$$

where $i = \sqrt{-1}$.

A detailed outline of the computational procedure is given below.

1. Read values for a, b, and c.
2. Print the values for a, b, and c (in order to identify the problem).
3. Determine if a = 0.
    (a) If a = 0,   calculate   x = -c/b.   Then print the value for x and stop.
    (b) If a ≠ 0,   continue with step 4 below.
4. Compare b² with 4ac.
    (a) If b² = 4ac,   calculate   x = -b/2a.   Then print the value for x and stop.
    (b) If   b² > 4ac,   calculate

$$x_1 = \frac{-b + \sqrt{b^2 - 4ac}}{2a}$$

$$x_2 = \frac{-b - \sqrt{b^2 - 4ac}}{2a}$$

Then print the values for x_1 and x_2, and stop.
    (c) If   b² < 4ac,   calculate

$$x_1 = \frac{-b + i\sqrt{4ac - b^2}}{2a}$$

$$x_2 = \frac{b \quad i\sqrt{4ac \quad b^2}}{2a}$$

Then print the values for x_1 and x_2, and stop.

Figure 7-10 shows a flowchart of the procedure.   (The capital letters enclosed in circles are flowchart connectors, which eliminate the need for cumbersome connecting lines.)   It would now be a relatively simple matter to write a detailed computer program from this flowchart (provided, of course, that one had adequate proficiency in a specific programming language), since the overall program structure has been completely designed.

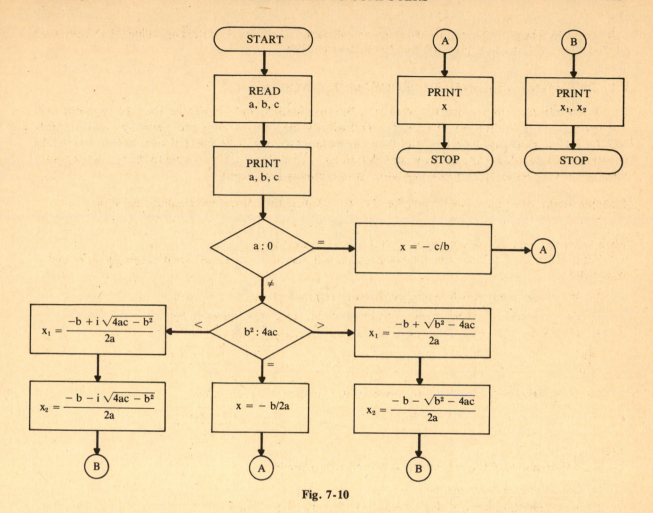

Fig. 7-10

It is important to understand that one need *not* be a skilled computer programmer in order to *design* a program. What is required is a clear view of the desired program features and an overall understanding of the computational methods that are required to implement those features. In fact, a computer program is often designed by the engineer or manager in charge of a project, whereas the detailed programming is usually carried out by a programming specialist.

Example 7.11. Design a computer program that will calculate the average of a list of numbers, i.e.

$$x_{avg} = \frac{x_1 + x_2 + \cdots + x_n}{n}$$

and will then calculate the deviation of each individual number about the average, using the expression

$$d_i = x_i - x_{avg} \qquad (i = 1, 2, ..., n)$$

This program will require two loops—one to calculate the desired average (see Example 7.7), and the other to calculate the individual deviations after the average has been obtained. Moreover, it will be necessary to store all the individual x's within the computer's memory so that they will be available for the computation of the d's.

A detailed outline of the computational procedure is presented below.

1. Set SUM = 0.
2. Carry out the following looping operation for I = 1, 2, ..., n.
    (*a*) Read and store a value for x_i.
    (*b*) Let SUM = SUM + x_i.
3. Calculate the desired average as x_{avg} = SUM/n.

4. Print the calculated value for x_{avg}.

5. Carry out the following looping operation for I = 1, 2, ..., n.

    (*a*)   Let $d_i = x_i - x_{avg}$.

    (*b*)   Print the current values for I, x_i, and d_i.

6. Stop.

Figure 7-11 shows a flowchart corresponding to the above outline.

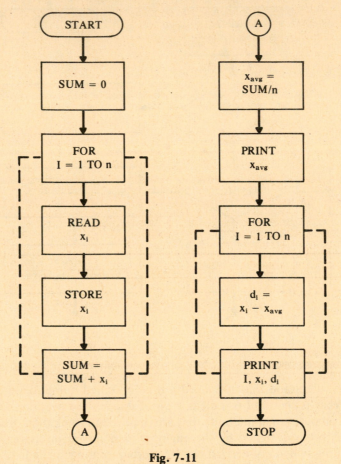

Fig. 7-11

Example 7.12. Design a computer program that will simulate the movement of a rubber ball as it bounces up and down under the force of gravity, while at the same time traveling in the horizontal direction with a constant velocity. The initial vertical displacement (the original height above the ground) will be specified (h), along with the horizontal velocity (u) and the number of times the ball bounces (n). Also known will be the bounce coefficient (c), which is the ratio of the vertical speed just after impact to the vertical speed just before impact, and the acceleration of gravity ($g = 9.80$ m/s²).

In order to determine the position of the ball at various times, we will make use of the following well-known equations of motion:

$$\frac{dx}{dt} = u \qquad \frac{dy}{dt} = v \qquad \frac{dv}{dt} = -g$$

where x is the horizontal displacement (originally zero at the start of the problem), y is the height above the ground (originally equal to h), v is the vertical velocity (originally zero), and t represents time.

A computer solution requires that the above equations be rewritten in terms of some finite time interval, Δt. Thus,

$$x_{i+1} = x_i + u\,\Delta t \tag{1}$$

$$v_{i+1} = v_i - g\,\Delta t \tag{2}$$

$$y_{i+1} = y_i + \frac{1}{2}(v_{i+1} + v_i)\,\Delta t \tag{3}$$

where the subscripts i and $i + 1$ refer to the values of the different variables at the start and end of the time increment, respectively. Moreover, the time can be advanced from one time increment to the next using the expression

$$t_{i+1} = t_i + \Delta t \qquad (4)$$

The above equations can be used to calculate the position of the ball within each time increment during which a bounce does not occur. If a bounce *does* occur, then the above equations must be modified somewhat. A bounce condition can be identified by a negative calculated value for y_{i+1}, which is physically impossible. When this condition occurs we can recalculate v_{i+1} and y_{i+1} as follows. First calculate the time required for the ball to hit the ground, beginning with its position at the start of the ith time increment. If we call this time Δt^*, then, from simple proportionality,

$$\frac{\Delta t^*}{\Delta t} = \frac{y_i - 0}{y_i - y_{i+1}}$$

Hence

$$\Delta t^* = \frac{y_i}{y_i - y_{i+1}} \Delta t \qquad (5)$$

We can compute the vertical velocity immediately *before* impact as

$$v = v_i - g\,\Delta t^*$$

so that the vertical velocity immediately *after* impact will be

$$v^* = -c(v_i - g\,\Delta t^*) \qquad (6)$$

Now the vertical velocity at the end of the time increment will be

$$v_{i+1} = v^* - g(\Delta t - \Delta t^*) \qquad (7)$$

and the vertical displacement at the end of the time increment can be written as

$$y_{i+1} = \frac{1}{2}(v^* + v_{i+1})(\Delta t - \Delta t^*) \qquad (8)$$

We now have enough information at our disposal to write an outline of a complete computer program. Specifically,

1. Read values for h, u, n, c, g, and Δt.

2. Initialize all parameters:

$$
\begin{aligned}
I &= 0 \qquad \text{(I is the increment counter)}\\
J &= 0 \qquad \text{(J is the bounce counter)}\\
t_i &= 0\\
x_i &= 0\\
v_i &= 0\\
y_i &= h
\end{aligned}
$$

3. For each successive time increment ($I = 1, 2, 3, \ldots$), proceed as follows:

   (*a*) Advance the time and then compute the horizontal displacement, the vertical velocity, and the vertical displacement for each successive time increment, using equations (*1*) through (*4*). Continue until the required number of bounces has occurred, i.e. until $J = n$.

   (*b*) If the ball hits the ground (as indicated by a negative value for y_{i+1}), test to see whether this is a bounce condition or a program termination.

   　(*i*) *Bounce condition* ($J < n$): Recalculate the vertical velocity and the vertical displacement using equations (*5*) through (*8*). Then advance the bounce counter ($J = J + 1$) and proceed to step 3(*c*) below.

   　(*ii*) *Terminal condition* ($J = n$): Determine the impact time using equation (*5*), and the corresponding horizontal displacement from (*1*). Then print the final values for t and x, and stop.

   (*c*) Print the calculated values for t_{i+1}, x_{i+1}, v_{i+1}, and y_{i+1}.

A corresponding flowchart is shown in Fig. 7-12. Notice that all newly calculated values (t_{i+1}, x_{i+1}, v_{i+1}, and y_{i+1}) are shown with the subscript i rather than i+1. This is because we have advanced the counter (I = I+1) at the start of each time increment; thus, the subscript i now refers to the values at the end of the current time increment.

From this flowchart it is relatively simple to construct a complete, detailed computer program in some technically oriented language such as FORTRAN or BASIC. Examples of such programs will be seen in Section 7.5.

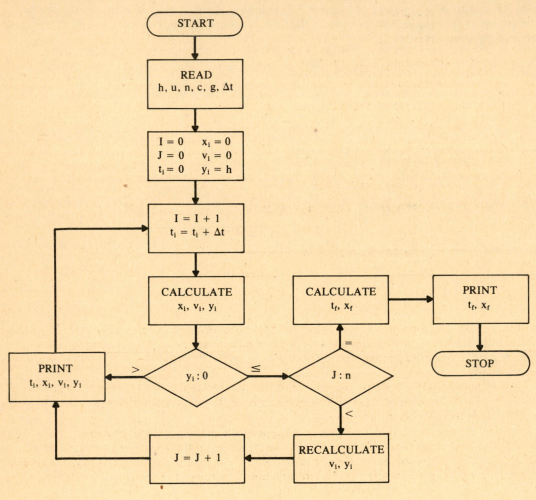

Fig. 7-12

7.5 PROGRAMMING LANGUAGES

In this book we will consider only general purpose, technically oriented languages. Thus, the commercially oriented COBOL will be omitted, despite its being the most widely used of all programming languages.

FORTRAN

FORTRAN may be used with all large computers and most mid-sized (mini) computers. FORTRAN consists largely of a mixture of algebraic-type expressions, augmented by certain English keywords, such as DO, IF, GO TO, READ, and WRITE. Thus a typical FORTRAN program will bear close resemblance to the manner in which an engineer would naturally outline the solution to a problem.   In fact, many engineers prefer to write a problem statement in FORTRAN rather than in more conventional mathematical terms.

Example 7.13.   Figure 7-13 shows a FORTRAN program that computes the average of a list of numbers and then calculates the deviation of each individual number about the average (see Example 7.11).

```
C FORTRAN PROGRAM TO COMPUTE THE AVERAGE OF A LIST OF NUMBERS AND
C THEN CALCULATE THE DEVIATION OF EACH NUMBER ABOUT THE AVERAGE
C
      DIMENSION X(100)
  100 FORMAT(I3)
  200 FORMAT(F10.5)
  300 FORMAT("1",10X,"THE AVERAGE VALUE IS XAVG=",F10.5//,12X,"I",12X,
     1 "X(I)",14X,"D(I)"/)
  400 FORMAT(11X,I3,2(8X,F10.5))
      READ (5,100) N
C
C CALCULATE AVERAGE VALUE OF THE X'S
C
      SUM=0.
      DO 1 I=1,N
      READ (5,200) X(I)
    1 SUM=SUM+X(I)
      XAVG=SUM/N
      WRITE (6,300) XAVG
C
C CALCULATE DEVIATIONS ABOUT THE AVERAGE
C
      DO 2 I=1,N
      D=X(I)-XAVG
    2 WRITE (6,400) I,X(I),D
      STOP
      END
```

Fig. 7-13

The lines preceded with the letter C are *comments*.   All the remaining lines represent FORTRAN *statements* (instructions), which cause specific actions to be taken by the computer.   Each line represents a separate instruction (except for the line between 300 FORMAT . . . and 400 FORMAT . . . , which is a continuation of the previous line).

The first statement (DIMENSION) reserves 100 words of storage within the computer's memory for the x's; thus the program can be used to average as many as 100 different x-values.   (Note that the program can average *fewer* than 100 x-values if desired.)

The next four statements (FORMAT) specify the appearance of the input and the output data, and provide English labels for the output data.   (A more precise interpretation of these statements is somewhat technical in nature, and beyond the scope of our present discussion.)   The next statement (READ) causes a value for n to be read into the computer, where n is the actual number of x-values to be averaged.   (Note that n cannot exceed 100.)

The desired value for x_{avg} is determined by the next six statements (ranging from   SUM = 0   to WRITE). The first of these statements initializes the cumulative sum at zero.   The following three statements (DO, READ, and   SUM = ...) constitute a loop that reads and stores all the x's, and calculates their cumulative sum.   After the loop has been executed n times, a value for x_{avg} is obtained (XAVG = ...) and then printed out (WRITE).

The last block of statements consists of another loop (DO,   D = ...,   and WRITE) that calculates and prints a value for each of the deviations.   (Actually, each pass through the loop causes the current values for i, x_i, and d_i to be printed.)   After this loop is a statement that causes the execution to cease (STOP), followed by a statement that identifies the physical end of the program (END).

The reader should note the close correspondence between the actual FORTRAN program shown in Fig. 7-13 and the flowchart presented in Fig. 7-11.

FORTRAN is an easy language to learn and to use.   It is used extensively by practicing engineers, and is learned by most engineering students at the freshman or sophomore level.   Many students now learn FORTRAN while they are still in high school.

BASIC

BASIC was originally intended as a simple timesharing language that could easily be used by high-school and college students having a minimum of formal training.   The language is now available at most large computer installations and through practically all commercial timesharing services.   Moreover, BASIC is now used with most mini- and microcomputers as the primary high-level programming language.

In many respects BASIC is similar to FORTRAN, though it is somewhat easier to learn and to use.   Thus the language has an algebra-like structure, which is augmented by certain English keywords, such as LET, IF, GO TO, INPUT, and PRINT.

Example 7.14.   Figure 7-14 shows a BASIC program that has the same purpose as the FORTRAN program shown in Fig. 7-13, i.e. the program computes the average of a list of numbers and then calculates the deviation of each individual number about the average (see Example 7.11).

```
>10    REM BASIC PROGRAM TO COMPUTE THE AVERAGE OF A LIST OF NUMBERS AND
>20    REM THEN CALCULATE THE DEVIATION OF EACH NUMBER ABOUT THE AVERAGE
>25
>30    DIM X(100)
>40    INPUT N
>45
>50    LET S=0                    "CALCULATE AVERAGE VALUE OF THE X'S
>60    FOR I=1 TO N
>70       INPUT X(I)
>80       LET S=S+X(I)
>90    NEXT I
>100   LET A=S/N
>110   PRINT "THE AVERAGE VALUE IS XAVG=";A
>115
>120   FOR I=1 TO N               "CALCULATE DEVIATIONS ABOUT THE AVERAGE
>130      LET D=X(I)-A
>140      PRINT "I=";I,"X(I)=";X(I),"D(I)=";D
>150   NEXT I
>160   END
```

Fig. 7-14

We see that every line in a BASIC program must begin with a *line number*.   Each of these lines, except the blank lines (lines 25, 45, and 115), represents a BASIC *statement* (instruction).

The first two lines are remarks.   These statements serve no purpose other than providing a program heading.   Line 30 causes 100 words of storage to be reserved for storing the x's within the computer's memory.   The next line (line 40) causes the current value for n (i.e. the actual number of x's that will be averaged) to be entered into the computer.   (Note that the value for n cannot exceed 100.)

Lines 50 through 90 cause the cumulative sum of the x's to be calculated.   (Notice that lines 60 through 90 define a loop which is executed n times, i.e. for I = 1,   I = 2,   ...,   I = N).   The desired value for x_{avg} is determined in line 100, and printed out (along with an appropriate label) in line 110.

A second loop is created by lines 120 through 150.   This loop calculates and prints a value for each of the deviations.   (Notice that the PRINT statement actually causes the current values for i, x_i, and d_i to be printed, with appropriate labels.)   Finally, the last line identifies the end of the program, thus causing the computation to cease.

The reader should note the close correspondence between the BASIC program presented in this example and the flowchart shown in Fig. 7-11.   Also, the reader should observe the similarity between this program and the corresponding FORTRAN program shown in Fig. 7-13.

Most engineering students are exposed to FORTRAN rather than BASIC while in college, although many students now learn BASIC before entering college.   The language is becoming increasingly popular, and may eventually rival FORTRAN as timesharing and microcomputer applications become more prevalent.

Other Technical Programming Languages

ALGOL, a FORTRAN-like language, is now available for use on many large computers. The language is in some respects superior to FORTRAN, although it never had FORTRAN's broad-based market appeal. The use of ALGOL now appears to be more prevalent in Europe than in the United States.

PL/1 is a "second-generation" programming language that was developed by IBM during the 1960's. This is a very comprehensive language that encompasses some of the best features of FORTRAN, ALGOL, and COBOL. Within the engineering community, however, the language has not received the widespread acceptance that had been expected, and the future of PL/1 is now somewhat doubtful. Nevertheless, PL/1 is one of the most powerful programming languages available for serious scientific and technical applications.

APL is a more advanced programming language that contains features not found in FORTRAN, BASIC, etc. In the past the language has been used only by a relatively small number of engineers, scientists, and mathematicians confronted with rather specialized, complex programming applications. A broader interest in APL appears to be developing, however, as the language becomes increasingly available.

Solved Problems

7.1. Modify the procedure presented in Example 7.1 so that a *sequence* of radii can be processed in succession (without stopping and restarting the program after every calculation).

Sequential program execution can easily be accomplished by means of a loop structure. One way to carry out the looping operation is simply to continue reading input data until the data have all been read. The computer will stop automatically once it has run out of data. This procedure, which is illustrated in Fig. 7-15(*a*), is especially suited to batch processing.

A more sophisticated approach is to assign a fictitious value for the radius, such as $r = 0$, which will indicate a stopping condition. Thus, the program will continue to execute until a value of $r = 0$ has been read, at which time the execution will cease. This method is particularly suitable for timesharing applications, though it can also be used for batch processing. Figure 7-15(*b*) illustrates this procedure.

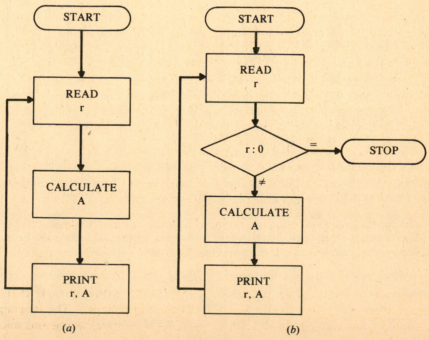

(*a*)   (*b*)

Fig. 7-15

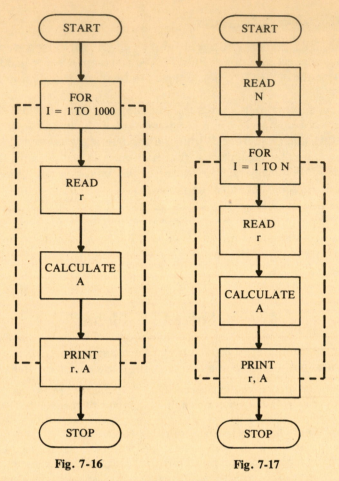

Fig. 7-16   **Fig. 7-17**

7.2.   Modify the procedure presented in Problem 7.1 so that a sequence of 1000 radii can be processed sequentially, using batch processing, as described in Example 7.2.

Figure 7-16 indicates the type of loop structure that is appropriate for this problem.

7.3.   Modify the procedure presented in Problem 7.2 so that a sequence of N radii can be processed sequentially, using batch processing, as described in Example 7.2.

The desired procedure is shown in Fig. 7-17. Notice that a numerical value for N is first read into the computer, followed by the loop that calculates the areas for the N different radii.

It is interesting to observe that this procedure is very well suited for batch processing, whereas the procedure shown in Fig. 7-15(b) is most appropriate for timesharing. Each of these procedures can, however, be used in either mode.

7.4.   Modify the procedure presented in Problem 7.3 so that (1) all the values for the radii are stored within the computer's memory as they are read in; (2) the corresponding areas are stored within the computer's memory as they are calculated; (3) all the areas are averaged; (4) the deviation of each calculated area about the average is determined and printed out.

The entire procedure is shown in Fig. 7-18. Notice that *two* loops are now required—one loop to calculate the average area, and the other to determine the deviations about the average.

7.5.   If a rocket is fired from a horizontal surface, its trajectory can be approximated by the formulas

$$x = (v_0 \cos \theta)t \qquad y = (v_0 \sin \theta)t - \frac{gt^2}{2}$$

where x = horizontal displacement at time t
y = vertical displacement at time t

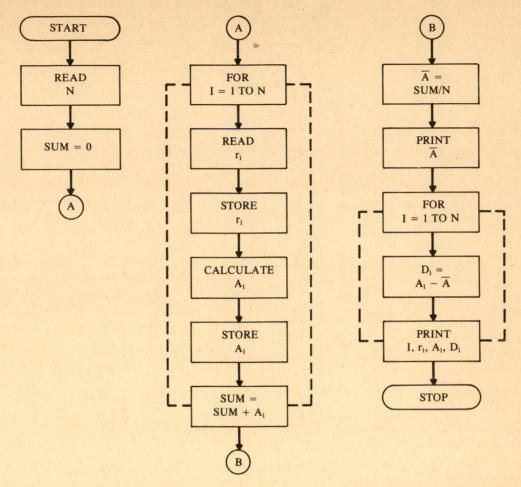

Fig. 7-18

$$v_0 = \text{initial speed of the rocket}$$
$$\theta = \text{launching inclination from the horizontal}$$
$$g = \text{gravitational acceleration}$$

Design a computer program that will accept v_0 and θ as input parameters and will then calculate t_f (the time required for the rocket to land) and x_f (the distance traveled). Include a provision for carrying out several successive calculations for different values of v_0 and θ. Let $v_0 = 0$ denote a stopping condition.

The time required for the rocket to land can be determined by setting $y = 0$ and solving for $t > 0$. Thus,

$$t_f = \frac{2v_0 \sin \theta}{g}$$

The distance traveled can now be determined as

$$x_f = (v_0 \cos \theta) \, t_f$$

The overall computational procedure is as follows:

1. Read values for v_0 and θ.
2. If $v_0 = 0$, stop; otherwise, continue below.
3. Calculate t_f.
4. Calculate x_f.
5. Print the values for v_0, θ, x_f, and t_f.
6. Go to step 1.

Figure 7-19 shows a corresponding flowchart.

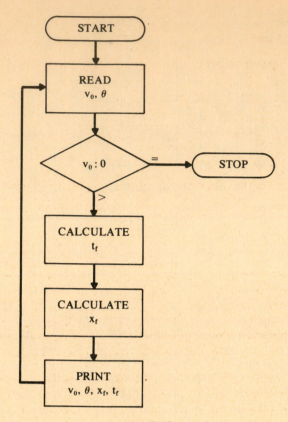

Fig. 7-19

7.6.   Figure 7-20 shows a FORTRAN program for the rocket trajectory of Problem 7.5.   Explain
the purpose of each line.

```
C FORTRAN PROGRAM TO CALCULATE TIME AND DISTANCE OF ROCKET FLIGHT
C
  100 FORMAT(2F7.2)
  200 FORMAT("0",10X,"INITIAL VELOCITY=",F7.2," METERS/SEC",10X,
      1 "ANGLE OF INCLINATION=",F7.2," RADIANS"///,10X,"TIME OF FLIGHT=",
      2 F7.2," SECONDS",10X,"DISTANCE TRAVELED=",F7.2," METERS"/)
        GRAV=9.807
      1 READ (5,100) VO,THETA
        IF (VO.EQ.0.) STOP
        TF=2.*VO*SIN(THETA)/GRAV
        XF=VO*COS(THETA)*TF
        WRITE (6,200) VO,THETA,TF,XF
        GO TO 1
        END
```

Fig. 7-20

The first two lines (each beginning with a C) are comments.   The next four lines are FORMAT
statements, which specify the appearance of the input and the output data and provide English labels
for the output.   A value for the gravitational acceleration is then assigned (GRAV = 9.807), and the
input data are read into the computer (READ).

A test (IF ...) is then carried out to see if a zero value has been specified for v_0. If so,
the computation terminates.   Otherwise, the computation proceeds to evaluate t_f (TF = ...) and
x_f (XF = ...).

Finally, the calculated values are printed out, along with the input data (in order to identify each
problem).   The computation then loops back to statement number 1 (READ), thus causing a new set
of input data to be read, etc.   The last statement (END) identifies the physical end of the program.

7.7. Figure 7-21 shows a BASIC program for the rocket trajectory of Problem 7.5. Explain the purpose of each line.

```
>10   REM BASIC PROGRAM TO CALCULATE TIME AND DISTANCE OF ROCKET FLIGHT
>20
>30   LET G=9.807
>40   PRINT "INITIAL VELOCITY (M/SEC) =";
>50   INPUT VO
>60   IF VO=0 GO TO 150
>70   PRINT "ANGLE OF INCLINATION (RADIANS) =";
>80   INPUT A
>90   LET T=2*VO*SIN(A)/G
>100  LET X=VO*COS(A)*T
>110  PRINT
>120  PRINT "TIME OF FLIGHT=";T;"SECONDS","DISTANCE TRAVELED=";X;"METERS"
>130
>140  GO TO 40
>150  END
```

Fig. 7-21

The first line is a remark (program heading), followed by a blank line. Line 30 assigns a value to g, the gravitational acceleration. Line 40 prints a message requesting information, and line 50 causes the requested information (the value for v_0) to be read into the computer.

A stopping condition is provided in line 60. If a zero value has been entered for v_0, then the computation jumps to line 150, which is the end of the program. Otherwise, the computation continues by printing a request for more information (line 70). Line 80 causes the requested information (the value for θ, which is now called A) to be read into the computer.

The desired values for t_f and x_f are calculated in lines 90 and 100, respectively. Line 120 causes these values to be printed out, with appropriate labels. This printed output is preceded and followed by blank lines of output (generated by lines 110 and 130) in order to double-space the output data. The computation then loops back to line 40, thus causing a new set of input data to be read, etc. Finally, line 150 provides a stopping instruction and indicates the end of the program.

Supplementary Problems

Answers to most problems are provided at the end of the book.

7.8. Design a computer program that will calculate the area of a rectangle whose length and width are given. Include a provision for batch-type repetitive program execution, so that N sets of data can be processed in succession (where N is an input parameter).

7.9. Design a computer program that will calculate the volume of a sphere whose radius is given. Include a provision for repetitive program execution in the timesharing mode (i.e. allow the program to continue to execute until a zero value for the radius has been encountered).

7.10. Design a computer program to solve each of the problems described below. For each problem, include a provision for batch-type repetitive program execution, so that N sets of input data can be processed in succession (where N is an input parameter).

(a) Convert a temperature reading in degrees Fahrenheit to degrees Celsius, using the formula

$$\theta_C = \frac{5}{9}(\theta_F - 32)$$

(b) A piggy bank contains n_1 half-dollars, n_2 quarters, n_3 dimes, n_4 nickels, and n_5 pennies. Determine how much money is in the bank, in terms of dollars.

(c) If a, b, and c represent the three sides of a triangle, then the area of the triangle is given by

$$A = \sqrt{s(s - a)(s - b)(s - c)}$$

where $s = (a + b + c)/2$. Also, the radius of the *largest inscribed circle* is given by

$$r_i = \frac{A}{s}$$

and the radius of the *smallest circumscribed circle* is

$$r_c = \frac{abc}{4A}$$

Calculate the area of the triangle, the area of the largest inscribed circle and the area of the smallest circumscribed circle, for a particular data set. (Note that a set of values a, b, and c will represent a triangle only if the largest value is smaller than the sum of the two remaining values.   Include a test for this condition within the program.)

7.11.   Design a computer program to solve each of the problems described below.   For each problem, include a provision for repetitive program execution in the timesharing mode.

(a)   Convert a temperature reading in degrees Fahrenheit to degrees Celsius, using the formula

$$\theta_C = \frac{5}{9}(\theta_F - 32)$$

(b)   The pressure, volume and temperature of a mass of air are related by the formula

$$pV = 2.83\, m(T + 273)$$

where p = pressure, atm
          V = volume, liters
          m = mass of air, kg
          T = temperature, °C

Determine the amount of air that is present in a container if the pressure, volume, and temperature are specified.

(c)   Calculate the future worth of a given sum of money, using the formula

$$F = P(1 + i)^n$$

where F = future worth
          P = given sum
          i = annual interest rate (expressed as a decimal)
          n = number of years

7.12.   The *Fibonacci numbers* are an interesting sequence in which each number is equal to the sum of the previous two numbers.   In other words,

$$F_i = F_{i-1} + F_{i-2}$$

where F_i is the ith Fibonacci number.   The first two Fibonacci numbers are defined to equal 1, i.e.

$$F_1 = F_2 = 1$$

Design a program that will generate and print out the first N Fibonacci numbers, where N is an input parameter.

7.13.   Design a program that will calculate the sum of the first N odd integers.   Enter the value for N as an input parameter.

7.14.   Design a program that will calculate the sum of the first N multiples of an integer K.   Enter the values for N and K as input parameters.

7.15. Design a program that will calculate $K!$, where K is a positive integer.   (Recall that $K!$ is defined as $1 \times 2 \times 3 \times \cdots \times K$.)   Enter the value for K as an input parameter.

7.16. Design a program that will calculate $\sin x$ by summing the first N terms of the series

$$\sin x = x - \frac{x^3}{3!} + \frac{x^5}{5!} - \frac{x^7}{7!} + \cdots \qquad (x \text{ in radians})$$

Enter the values for x and N as input parameters.

7.17. Design a program that will calculate $\sin x$ using the infinite series presented in Problem 7.16.   Rather than sum a specified number of terms, however, continue to add successive terms in the series until the value of a term becomes smaller (in magnitude) than 10^{-5}.

7.18. Design a program that will read in a list of numbers and then reorder the numbers from smallest to largest.   Include a provision for printing out both the original list and the reordered list.   Design the program in such a manner that only one list of numbers is stored within the computer's memory.

7.19. Design a program that will calculate the weighted average of a list of numbers.   Remember to read in both the numbers themselves and their corresponding weighting factors.   (See Section 4.3.)

7.20. Design a program that will calculate the mean and the standard deviation of a list of numbers.   (See Section 4.4.)

7.21. Design a program that will pass a straight line through a set of M data points using the method of least squares.   (See Section 4.9.)   Assume that a routine is available to solve the system of simultaneous linear equations (represent the solution as a single step in the overall computational procedure).

7.22. Design a program that will pass a power function through a set of M data points using the method of least squares.   (See Section 4.9.)

7.23. Design a program that will calculate the correlation coefficient for a set of M data points.   (See Section 4.10.)

7.24. Design a program that will determine the end-of-semester QPA for a college student.   (See Problem 4.9.)

7.25. Design programs that will obtain a solution to a nonlinear algebraic equation using (a) the method of successive substitution, (b) the Newton-Raphson method, (c) the method of bisection.   (See Section 5.2.)   In each case read appropriate values for the initial guess (x_0) and the stopping criterion (ϵ) into the computer.   In (a) and (b) make allowance for possible divergence of the method.

7.26. Design programs that will obtain a solution to a system of simultaneous linear equations using (a) Gauss-Jordan elimination, (b) Gauss elimination. (See Section 5.3.) *Hint*: Each program will require a set of *three* nested loops.

7.27. Design a program that will determine the area under the curve defined by the data points (x_1, y_1), (x_2, y_2), $\ldots$, (x_n, y_n), using the trapezoidal rule.   (See Section 5.5.)

7.28. Design a program that will evaluate the integral

$$I = \int_a^b y \, dx$$

using   (a)   the trapezoidal rule,   (b)   Simpson's rule, for a known integrand $y = f(x)$.   (See Section 5.5.)   In each case assume an even number of equal subintervals.   Enter the subinterval width (Δx) as an input parameter.

7.29. Design a program that will generate a table giving the equal-annual-payment, compound-interest factor (F/A) as a function of interest rate (i) and number of years (n).   Let $n = 1, 2, 3, \ldots, 30$ and let $i = 0.04, 0.05, 0.06, \ldots, 0.12$.   (See Section 6.2.)

7.30. Figure 7-22 contains a FORTRAN program for obtaining the roots of a quadratic equation, as described in Example 7.10.   Describe, as best you can, the purpose of each line in the program.

```
C FORTRAN PROGRAM TO COMPUTE THE ROOTS OF A QUADRATIC EQUATION
C
  100 FORMAT(I3)
  200 FORMAT(3F10.4)
  300 FORMAT('0',10X,'A=',F10.4,10X,'B=',F10.4,10X,'C=',F10.4)
  400 FORMAT('0',10X,'REAL ROOTS:',10X,'X1=',F10.4,10X,'X2=',F10.4//)
  500 FORMAT('0',10X,'REPEATED SINGLE ROOT:',10X,'X=',F10.4//)
  600 FORMAT('0',10X,'COMPLEX ROOTS:',10X,'X1=',F10.4,' + ',F10.4,'I',
    1 10X,'X2=',F10.4,' - ',F10.4,'I',//)
  700 FORMAT('0',10X,'LINEAR CASE - SINGLE ROOT:',10X,'X=',F10.4//)
      READ (5,100) N
      DO 5 I=1,N
      READ (5,200) A,B,C
      WRITE (6,300) A,B,C
      IF (A.EQ.0.) GO TO 4
      DISC=B**2-4.*A*C
      IF (DISC) 3,2,1
C
C REAL ROOTS (DISC>0)
C
    1 X1=(-B+SQRT(DISC))/(2.*A)
      X2=(-B-SQRT(DISC))/(2.*A)
      WRITE (6,400) X1,X2
      GO TO 5
C
C REPEATED ROOT (DISC=0)
C
    2 X=-B/(2.*A)
      WRITE (6,500) X
      GO TO 5
C
C COMPLEX ROOTS (DISC<0)
C
    3 XREAL=-B/(2.*A)
      XIMAG=SQRT(-DISC)/(2.*A)
      WRITE (6,600) XREAL,XIMAG,XREAL,XIMAG
      GO TO 5
C
C LINEAR CASE - SINGLE ROOT (A=0)
    4 X=-C/B
      WRITE (6,700) X
    5 CONTINUE
      STOP
      END
```

Fig. 7-22

7.31.		Figure 7-23 contains a BASIC program for obtaining the roots of a quadratic equation, as described
		in Example 7.10. Describe, as best you can, the purpose of each line in the program.

```
>10  REM BASIC PROGRAM TO COMPUTE THE ROOTS OF A QUADRATIC EQUATION
>20
>30  PRINT "VALUES FOR A, B AND C";
>40  INPUT A,B,C
>50  IF (A+B+C)=0 THEN 500
>60  IF A=0 THEN 430
>70  LET D=B↑2-4*A*C
>80  IF D<0 THEN 330
>90  IF D=0 THEN 230
>100
>110 REM REAL ROOTS (D>0)
>120
>130 LET X1=(-B+SQR(D))/(2*A)
>140 LET X2=(-B-SQR(D))/(2*A)
>150 PRINT "REAL ROOTS:","X1=";X1,"X2=";X2
>160 GO TO 30
>200
>210 REM REPEATED ROOT (D=0)
>220
>230 LET X=-B/(2*A)
>240 PRINT "REPEATED SINGLE ROOT:","X=";X
>250 GO TO 30
>300
>310 REM COMPLEX ROOTS (D<0)
>320
>330 LET R=-B/(2*A)
>340 LET I=SQR(-D)/(2*A)
>350 PRINT "COMPLEX ROOTS:","X1=";R;" + ";I,"I","X2=";R;" - ";I;"I"
>360 GO TO 30
>400
>410 REM LINEAR CASE - SINGLE ROOT (A=0)
>420
>430 LET X=-C/B
>440 PRINT "LINEAR CASE - SINGLE ROOT:","X=";X
>450 GO TO 30
>500 END
```

					Fig. 7-23

Appendix A

Units Equivalences

Dimension	Units Equivalences

Area

$1 \text{ acre} = 43\,560 \text{ ft}^2 = 4.046\,856 \times 10^{-3} \text{ km}^2 = 1.5625 \times 10^{-3} \text{ mi}^2$

Density

$1 \text{ g/cm}^3 = 1000 \text{ kg/m}^3 = 62.428 \text{ lb}_m/\text{ft}^3$

$1 \text{ kg/m}^3 = 10^{-3} \text{ g/cm}^3 = 0.062\,428 \text{ lb}_m/\text{ft}^3$

$1 \text{ lb}_m/\text{ft}^3 = 0.016\,018\,5 \text{ g/cm}^3 = 16.0185 \text{ kg/m}^3$

Energy

$1 \text{ Btu} = 777.649 \text{ ft} \cdot \text{lb}_f = 1054.35 \text{ J} = 0.251\,996 \text{ kcal}$

$1 \text{ ft} \cdot \text{lb}_f = 1.285\,93 \times 10^{-3} \text{ Btu} = 1.355\,82 \text{ J}$

$1 \text{ hp} \cdot \text{hr} = 2546.14 \text{ Btu} = 1.98 \times 10^6 \text{ ft} \cdot \text{lb}_f = 2.684\,52 \times 10^6 \text{ J}$

$1 \text{ J} = 0.737\,562 \text{ ft} \cdot \text{lb}_f = 2.390\,06 \times 10^{-4} \text{ kcal} = 0.277\,778 \times 10^{-6} \text{ kW} \cdot \text{hr}$

$1 \text{ kcal} = 3.968\,32 \text{ Btu} = 3085.96 \text{ ft} \cdot \text{lb}_f = 4184 \text{ J}$

$1 \text{ kW} \cdot \text{hr} = 3409.52 \text{ Btu} = 2.655\,22 \times 10^6 \text{ ft} \cdot \text{lb}_f = 3.6 \times 10^6 \text{ J}$

Force

$1 \text{ dyne} = 10^{-5} \text{ N} = 7.233 \times 10^{-5} \text{ poundal}$

$1 \text{ lb}_f = 4.448\,22 \text{ N} = 32.174 \text{ poundals}$

$1 \text{ N} = 0.224\,809 \text{ lb}_f = 7.233 \text{ poundals}$

$1 \text{ poundal} = 0.031\,081 \text{ lb}_f = 0.138\,255 \text{ N}$

Heat capacity

$1 \text{ Btu/lb}_m \cdot {}^\circ\text{F} = 777.649 \text{ ft} \cdot \text{lb}_f/\text{lb}_m \cdot {}^\circ\text{R} = 1 \text{ kcal/kg} \cdot {}^\circ\text{C}$

$1 \text{ ft} \cdot \text{lb}_f/\text{lb}_m \cdot {}^\circ\text{R} = 1.285\,93 \times 10^{-3} \text{ Btu/lb}_m \cdot {}^\circ\text{F} = 1.285\,93 \times 10^{-3} \text{ kcal/kg} \cdot {}^\circ\text{C}$

$1 \text{ J/kg} \cdot \text{K} = 2.390\,06 \times 10^{-4} \text{ Btu/lb}_m \cdot {}^\circ\text{F} = 2.390\,06 \times 10^{-4} \text{ kcal/kg} \cdot {}^\circ\text{C}$

$1 \text{ kcal/kg} \cdot {}^\circ\text{C} = 1 \text{ Btu/lb}_m \cdot {}^\circ\text{F} = 4184 \text{ J/kg} \cdot \text{K}$

Length

$1 \text{ angstrom (Å)} = 10^{-10} \text{ m}$ $1 \text{ m} = 3.280\,84 \text{ ft} = 39.37 \text{ in}$

$1 \text{ cm} = 0.01 \text{ m}$ $1 \text{ micron } (\mu) = 10^{-6} \text{ m}$

$1 \text{ ft} = 12 \text{ in} = 0.3048 \text{ m}$ $1 \text{ mi} = 5280 \text{ ft} = 1.609\,344 \text{ km} = 1760 \text{ yd}$

$1 \text{ in} = 2.54 \text{ cm} = 0.083\,333\,3 \text{ ft}$ $1 \text{ mm} = 10^{-3} \text{ m}$

$1 \text{ km} = 1000 \text{ m} = 0.621\,371 \text{ mi}$ $1 \text{ yd} = 3 \text{ ft} = 36 \text{ in}$

Mass

$1 \text{ g} = 10^{-3} \text{ kg}$ $1 \text{ slug} = 14.5939 \text{ kg} = 32.174 \text{ lb}_m$

$1 \text{ kg} = 1000 \text{ g} = 2.204\,62 \text{ lb}_m$ $1 \text{ ton} = 907.185 \text{ kg} = 2000 \text{ lb}_m$

$1 \text{ lb}_m = 0.453\,592 \text{ kg}$

Power

$1 \text{ Btu/hr} = 3.927\,52 \times 10^{-4} \text{ hp} = 0.292\,875 \text{ W}$

$1 \text{ hp} = 550 \text{ ft} \cdot \text{lb}_f/\text{sec} = 745.7 \text{ W}$

$1 \text{ kcal/s} = 5.610\,84 \text{ hp} = 4184 \text{ W}$

$1 \text{ kW} = 1.341\,02 \text{ hp} = 1000 \text{ W}$

$1 \text{ MW} = 10^6 \text{ W}$

$1 \text{ W} = 0.737\,562 \text{ ft} \cdot \text{lb}_f/\text{sec} = 1.341\,02 \times 10^{-3} \text{ hp}$

Pressure

1 atm = 29.9213 inHg = 2116.224 lb_f/ft^2 = 760 mmHg = 1.013 26 × 10^5 Pa = 14.696 psi

1 lb_f/ft^2 = 47.88 Pa = 6.944 44 × 10^{-3} psi

1 mmHg = 2.7845 lb_f/ft^2

1 N/cm^2 = 10^4 Pa

1 pascal (Pa) = 1 N/m^2 = 2.088 56 × 10^{-2} lb_f/ft^2

1 psi (lb_f/in^2) = 6894.8 Pa = 144 lb_f/ft^2

Temperature

1 °C = 1 K = 1.8 °F = 1.8 °R

1 °F = 1 °R = 0.555 556 °C = 0.555 556 K

Temperature conversion formulas:

$$\theta_C = \theta_K - 273.15$$
$$\theta_F = 1.8\,\theta_C + 32 = \theta_R - 459.67$$
$$\theta_R = 1.8\,\theta_K$$

Thermal conductivity

1 Btu/hr · ft · °F = 4.1364 × 10^{-4} kcal/s · m · °C = 1.7307 W/m · K

1 kcal/s · m · °C = 2417.56 Btu/hr · ft · °F = 4184 W/m · K

1 W/m · K = 2.390 06 × 10^{-4} kcal/s · m · °C

Time

1 day (d) = 24 h = 8.64 × 10^4 s

1 h = 60 min = 3600 s

1 min = 60 s

1 ms = 10^{-3} s

1 μs = 10^{-6} s

1 ns = 10^{-9} s

1 ps = 10^{-12} s

Viscosity

1 Pa · s = 1 kg/m · s = 2419.1 lb_m/ft · hr = 10 poises

1 poise = 1 g/cm · s = 241.91 lb_m/ft · hr = 0.067 197 lb_m/ft · sec = 0.1 Pa · s

1 centipoise = 2.4191 lb_m/ft · hr = 0.001 Pa · s = 0.01 poise

1 kg/m · s = 10 g/cm · s = 2419.1 lb_m/ft · hr = 0.671 97 lb_m/ft · sec = 1 Pa · s

1 lb_m/ft · hr = 0.413 377 centipoise = 4.133 77 × 10^{-4} kg/m · s

= 2.7778 × 10^{-4} lb_m/ft · sec = 4.133 77 × 10^{-4} Pa · s = 4.133 77 × 10^{-3} poise

1 lb_m/ft · sec = 1488.16 centipoises = 3600 lb_m/ft · hr = 1.488 16 Pa · s = 14.8816 poises

Viscosity (kinematic)

1 centistokes = 10^{-6} m^2/s = 0.01 stokes

1 ft^2/hr = 2.7778 × 10^{-4} ft^2/sec = 0.258 064 stokes

1 ft^2/sec = 3600 ft^2/hr = 929.03 stokeses

1 m^2/s = 3.875 × 10^4 ft^2/hr = 10.7639 ft^2/sec = 10^4 stokeses

1 stokes = 1 cm^2/s = 10^{-4} m^2/s

Volume

1 gallon (gal) = 0.133 681 ft^3 = 3.785 31 L = 4 qt

1 liter (L) = 1000.028 cm^3 = 1.05672 qt

1 pint (pt) = 0.125 gal = 0.473 163 L = 0.5 qt

1 quart (qt) = 0.25 gal = 0.946 326 L = 2 pt

Single-Payment, Compound-Interest Factor

$$F/P = (1 + i)^n$$

n	i = 4%	5%	6%	7%	8%	9%	10%	12%	15%	20%
1	1.0400	1.0500	1.0600	1.0700	1.0800	1.0900	1.1000	1.1200	1.1500	1.2000
2	1.0816	1.1025	1.1236	1.1449	1.1664	1.1881	1.2100	1.2544	1.3225	1.4400
3	1.1249	1.1576	1.1910	1.2250	1.2597	1.2950	1.3310	1.4049	1.5209	1.7280
4	1.1699	1.2155	1.2625	1.3108	1.3605	1.4116	1.4641	1.5735	1.7490	2.0736
5	1.2167	1.2763	1.3382	1.4026	1.4693	1.5386	1.6105	1.7623	2.0114	2.4883
6	1.2653	1.3401	1.4185	1.5007	1.5869	1.6771	1.7716	1.9738	2.3131	2.9860
7	1.3159	1.4071	1.5036	1.6058	1.7138	1.8280	1.9487	2.2107	2.6600	3.5832
8	1.3686	1.4775	1.5938	1.7182	1.8509	1.9926	2.1436	2.4760	3.0590	4.2998
9	1.4233	1.5513	1.6895	1.8385	1.9990	2.1719	2.3579	2.7731	3.5179	5.1598
10	1.4802	1.6289	1.7908	1.9672	2.1589	2.3674	2.5937	3.1058	4.0456	6.1917
11	1.5395	1.7103	1.8983	2.1049	2.3316	2.5804	2.8531	3.4785	4.6524	7.4301
12	1.6010	1.7959	2.0122	2.2522	2.5182	2.8127	3.1384	3.8960	5.3503	8.9161
13	1.6651	1.8856	2.1329	2.4098	2.7196	3.0658	3.4523	4.3635	6.1528	10.6993
14	1.7317	1.9799	2.2609	2.5785	2.9372	3.3417	3.7975	4.8871	7.0757	12.8392
15	1.8009	2.0789	2.3966	2.7590	3.1722	3.6425	4.1772	5.4736	8.1371	15.4070
16	1.8730	2.1829	2.5404	2.9522	3.4259	3.9703	4.5950	6.1304	9.3576	18.4884
17	1.9479	2.2920	2.6928	3.1588	3.7000	4.3276	5.0545	6.8660	10.7613	22.1861
18	2.0258	2.4066	2.8543	3.3799	3.9960	4.7171	5.5599	7.6900	12.3755	26.6233
19	2.1068	2.5269	3.0256	3.6165	4.3157	5.1417	6.1159	8.6128	14.2318	31.9480
20	2.1911	2.6533	3.2071	3.8697	4.6610	5.6044	6.7275	9.6463	16.3665	38.3376
21	2.2788	2.7860	3.3996	4.1406	5.0338	6.1088	7.4002	10.8038	18.8215	46.0051
22	2.3699	2.9253	3.6035	4.4304	5.4365	6.6586	8.1403	12.1003	21.6447	55.2061
23	2.4647	3.0715	3.8197	4.7405	5.8715	7.2579	8.9543	13.5523	24.8915	66.2474
24	2.5633	3.2251	4.0489	5.0724	6.3412	7.9111	9.8497	15.1786	28.6252	79.4968
25	2.6658	3.3864	4.2919	5.4274	6.8485	8.6231	10.8347	17.0001	32.9190	95.3962
26	2.7725	3.5557	4.5494	5.8074	7.3964	9.3992	11.9182	19.0401	37.8568	114.4754
27	2.8834	3.7335	4.8223	6.2139	7.9881	10.2451	13.1100	21.3249	43.5353	137.3705
28	2.9987	3.9201	5.1117	6.6488	8.6271	11.1671	14.4210	23.8839	50.0656	164.8446
29	3.1187	4.1161	5.4184	7.1143	9.3173	12.1722	15.8631	26.7499	57.5755	197.8136
30	3.2434	4.3219	5.7435	7.6123	10.0627	13.2677	17.4494	29.9599	66.2118	237.3763

Appendix C

Equal-Annual-Payment, Compound-Interest Factor

$$F/A = \frac{(1 + i)^n - 1}{i}$$

n	i = 4%	5%	6%	7%	8%	9%	10%	12%	15%	20%
1	1.0000	1.0000	1.0000	1.0000	1.0000	1.0000	1.0000	1.0000	1.0000	1.0000
2	2.0400	2.0500	2.0600	2.0700	2.0800	2.0900	2.1000	2.1200	2.1500	2.2000
3	3.1216	3.1525	3.1836	3.2149	3.2464	3.2781	3.3100	3.3744	3.4725	3.6400
4	4.2465	4.3101	4.3746	4.4399	4.5061	4.5731	4.6410	4.7793	4.9934	5.3680
5	5.4163	5.5256	5.6371	5.7507	5.8666	5.9847	6.1051	6.3528	6.7424	7.4416
6	6.6330	6.8019	6.9753	7.1533	7.3359	7.5233	7.7156	8.1152	8.7537	9.9299
7	7.8983	8.1420	8.3938	8.6540	8.9228	9.2004	9.4872	10.0890	11.0668	12.9159
8	9.2142	9.5491	9.8975	10.2598	10.6366	11.0285	11.4359	12.2997	13.7268	16.4991
9	10.5828	11.0266	11.4913	11.9780	12.4876	13.0210	13.5795	14.7757	16.7858	20.7989
10	12.0061	12.5779	13.1808	13.8164	14.4866	15.1929	15.9374	17.5487	20.3037	25.9587
11	13.4864	14.2068	14.9716	15.7836	16.6455	17.5603	18.5312	20.6546	24.3493	32.1504
12	15.0258	15.9171	16.8699	17.8885	18.9771	20.1407	21.3843	24.1331	29.0017	39.5805
13	16.6268	17.7130	18.8821	20.1406	21.4953	22.9534	24.5227	28.0291	34.3519	48.4966
14	18.2919	19.5986	21.0151	22.5505	24.2149	26.0192	27.9750	32.3926	40.5047	59.1959
15	20.0236	21.5786	23.2760	25.1290	27.1521	29.3609	31.7725	37.2797	47.5804	72.0351
16	21.8245	23.6575	25.6725	27.8881	30.3243	33.0034	35.9497	42.7533	55.7175	87.4421
17	23.6975	25.8404	28.2129	30.8402	33.7502	36.9737	40.5447	48.8837	65.0751	105.9305
18	25.6454	28.1324	30.9056	33.9990	37.4502	41.3013	45.5992	55.7497	75.8364	128.1167
19	27.6712	30.5390	33.7600	37.3790	41.4463	46.0185	51.1591	63.4397	88.2118	154.7400
20	29.7781	33.0659	36.7856	40.9955	45.7620	51.1601	57.2750	72.0524	102.4436	186.6880
21	31.9692	35.7192	39.9927	44.8652	50.4229	56.7645	64.0025	81.6987	118.8101	225.0256
22	34.2480	38.5052	43.3923	49.0057	55.4567	62.8733	71.4027	92.5026	137.6316	271.0307
23	36.6179	41.4305	46.9958	53.4361	60.8933	69.5319	79.5430	104.6029	159.2764	326.2368
24	39.0826	44.5020	50.8156	58.1767	66.7647	76.7898	88.4973	118.1552	184.1678	392.4842
25	41.6459	47.7271	54.8645	63.2490	73.1059	84.7009	98.3470	133.3339	212.7930	471.9810
26	44.3117	51.1134	59.1564	68.6765	79.9544	93.3240	109.1817	150.3339	245.7120	567.3772
27	47.0842	54.6691	63.7058	74.4838	87.3507	102.7231	121.0999	169.3740	283.5688	681.8527
28	49.9676	58.4026	68.5281	80.6977	95.3388	112.9682	134.2099	190.6989	327.1041	819.2232
29	52.9663	62.3227	73.6398	87.3465	103.9659	124.1353	148.6309	214.5827	377.1697	984.0678
30	56.0849	66.4388	79.0582	94.4608	113.2832	136.3075	164.4940	241.3327	434.7451	1181.8814

Appendix D

Equal-Annual-Payment, Capital-Recovery Factor

$$A/P = \frac{i(1+i)^n}{(1+i)^n - 1}$$

n	i = 4%	5%	6%	7%	8%	9%	10%	12%	15%	20%
1	1.0400	1.0500	1.0600	1.0700	1.0800	1.0900	1.1000	1.1200	1.1500	1.2000
2	0.5302	0.5378	0.5454	0.5531	0.5608	0.5685	0.5762	0.5917	0.6151	0.6545
3	0.3603	0.3672	0.3741	0.3811	0.3880	0.3951	0.4021	0.4163	0.4380	0.4747
4	0.2755	0.2820	0.2886	0.2952	0.3019	0.3087	0.3155	0.3292	0.3503	0.3863
5	0.2246	0.2310	0.2374	0.2439	0.2505	0.2571	0.2638	0.2774	0.2983	0.3344
6	0.1908	0.1970	0.2034	0.2098	0.2163	0.2229	0.2296	0.2432	0.2642	0.3007
7	0.1666	0.1728	0.1791	0.1856	0.1921	0.1987	0.2054	0.2191	0.2404	0.2774
8	0.1485	0.1547	0.1610	0.1675	0.1740	0.1807	0.1874	0.2013	0.2229	0.2606
9	0.1345	0.1407	0.1470	0.1535	0.1601	0.1668	0.1736	0.1877	0.2096	0.2481
10	0.1233	0.1295	0.1359	0.1424	0.1490	0.1558	0.1627	0.1770	0.1993	0.2385
11	0.1141	0.1204	0.1268	0.1334	0.1401	0.1469	0.1540	0.1684	0.1911	0.2311
12	0.1066	0.1128	0.1193	0.1259	0.1327	0.1397	0.1468	0.1614	0.1845	0.2253
13	0.1001	0.1065	0.1130	0.1197	0.1265	0.1336	0.1408	0.1557	0.1791	0.2206
14	0.0947	0.1010	0.1076	0.1143	0.1213	0.1284	0.1357	0.1509	0.1747	0.2169
15	0.0899	0.0963	0.1030	0.1098	0.1168	0.1241	0.1315	0.1468	0.1710	0.2139
16	0.0858	0.0923	0.0990	0.1059	0.1130	0.1203	0.1278	0.1434	0.1679	0.2114
17	0.0822	0.0887	0.0954	0.1024	0.1096	0.1170	0.1247	0.1405	0.1654	0.2094
18	0.0790	0.0855	0.0924	0.0994	0.1067	0.1142	0.1219	0.1379	0.1632	0.2078
19	0.0761	0.0827	0.0896	0.0968	0.1041	0.1117	0.1195	0.1358	0.1613	0.2065
20	0.0736	0.0802	0.0872	0.0944	0.1019	0.1095	0.1175	0.1339	0.1598	0.2054
21	0.0713	0.0780	0.0850	0.0923	0.0998	0.1076	0.1156	0.1322	0.1584	0.2044
22	0.0692	0.0760	0.0830	0.0904	0.0980	0.1059	0.1140	0.1308	0.1573	0.2037
23	0.0673	0.0741	0.0813	0.0887	0.0964	0.1044	0.1126	0.1296	0.1563	0.2031
24	0.0656	0.0725	0.0797	0.0872	0.0950	0.1030	0.1113	0.1285	0.1554	0.2025
25	0.0640	0.0710	0.0782	0.0858	0.0937	0.1018	0.1102	0.1275	0.1547	0.2021
26	0.0626	0.0696	0.0769	0.0846	0.0925	0.1007	0.1092	0.1267	0.1541	0.2018
27	0.0612	0.0683	0.0757	0.0834	0.0914	0.0997	0.1083	0.1259	0.1535	0.2015
28	0.0600	0.0671	0.0746	0.0824	0.0905	0.0989	0.1075	0.1252	0.1531	0.2010
29	0.0589	0.0660	0.0736	0.0814	0.0896	0.0981	0.1067	0.1247	0.1527	0.2010
30	0.0578	0.0651	0.0726	0.0806	0.0888	0.0973	0.1061	0.1241	0.1523	0.2008

Answers to Supplementary Problems

Chapter 1

1.11.
(a) -4.214×10^{-1}
(e) -4.1403×10^{2}
(b) 5.0296×10^{-6}
(f) $1. \times 10^{-1}$ (or 1×10^{-1})
(c) 8.231×10^{1}
(g) 1.00000×10^{-1}
(d) -3.8651066×10^{7}
(h) -8.911×10^{-4}

1.12.
(a) $-0.4214E+0$
(c) $0.8231E+2$
(e) $-0.41403E+3$
(g) $0.100000E+0$
(b) $0.50296E-5$
(d) $-0.38651066E+8$
(f) $0.1E+0$
(h) $-0.8911E-3$

1.13.
(a) 0.2
(c) 0.0006776
(e) 0.0001748
(g) -0.005675
(i) 21050000
(b) -20
(d) -67.76
(f) 645100
(h) 0.0000001667
(j) -0.01

1.14.
(a) unknown
(c) one
(e) one
(g) four
(i) six
(b) three
(d) four
(f) three
(h) five
(j) six

1.15.
(a) 129.
(c) 0.999
(e) 1.00
(g) 1.01
(i) -0.113
(b) -0.00483
(d) 1.00
(f) 1.00
(h) 42600000
(j) 0.801

1.16.
(a) 128.
(c) 0.998
(e) 0.999
(g) 1.00
(i) -0.113
(b) -0.00483
(d) 0.999
(f) 1.00
(h) 42500000
(j) 0.800

1.17.
(a) 4.934×10^{-4}
(c) 5.268×10^{5}
(e) 3.647×10^{2}
(g) 4.524
(b) 2.842×10^{3}
(d) 8.737×10^{-6}
(f) -6.756×10^{-1}
(h) 609.4

1.18.
(a) 2.3
(b) 13
(c) 300
(d) 0.03

1.19.
(a) 0.3054
(b) -3.491
(c) 6.283
(d) 8.727

1.20.
(a) $5.730°$
(b) $171.9°$
(c) $-2.865°$
(d) $458.4°$

1.21.
(a) 1
(d) $-\infty$
(g) 0
(j) 1
(m) -1
(b) 0
(e) 0
(h) 1
(k) 0
(n) 1
(c) 0 (approx.)
(f) 0
(i) 0
(l) -1

1.22. $\sin(-x) = -\sin x$

1.23. $\cos(-x) = \cos x$

1.24. $\sin^2 x + \cos^2 x = 1$

1.25.
(a) 0.965
(c) 0.015
(e) 0.9993875
(b) 0.035
(d) -0.035
(f) 0.035

1.26. (a) 0.172 405 16   (c) 16.313 381   (e) 0.009 960 22
           (b) 0.317 510 27   (d) 3.979 428 8   (f) 0.809 676 09

1.27. (a) 1.213 mA   (b) 0.0821   (c) 3.794 s   (d) See Fig. A-1.

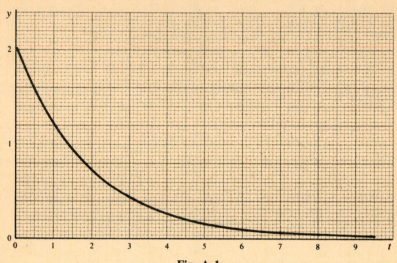

Fig. A-1

1.28. (a) 0.1837   (c) 25.06 h   (e) See Fig. A-2.
           (b) 0.4510   (d) 0.6059

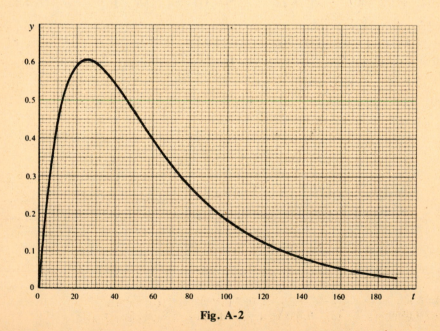

Fig. A-2

1.29. (a) 0.078 37 m   (b) 0.053 14 m/s   (c) See Fig. A-3.

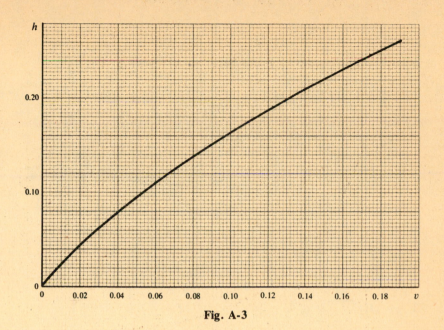

Fig. A-3

1.30. (a) 827.8 °C   (c) 100 °C   (e) See Fig. A-4.
(b) 159.7 °C   (d) 17.92 s

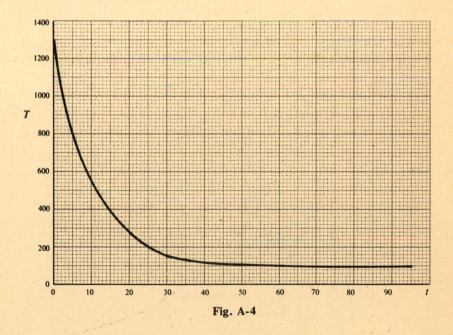

Fig. A-4

1.31. (a) 0 mV   (c) -8.584×10^{-7} mV   (e) 0 mV
(b) 0.0569 mV   (d) 0.088 46 mV, at 4.636 s   (f) See Fig. A-5.

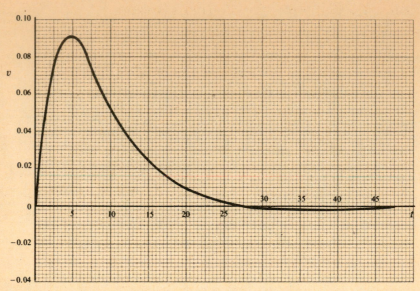

Fig. A-5

1.32. (a) RPN; $\sqrt{(1.47 + 0.86)(2.93 - 1.66)}$ (b) AOS; $-7.5e^{-(0.8)(3.88)/\pi}$

Chapter 2

2.23. $[L], [M], [T], [A], [\theta], [I]$

2.24. $[L], [M], [T], [F], [Q], [\theta], [I]$

2.25.
(a) $[FT]$	(c) $[FL]$	(e) $[FT^2/L^4]$	(g) $[F/T\theta]$	(i) $[FL/A^2T]$
(b) $[F/L^2]$	(d) $[FL/T]$	(f) $[FT/L^2]$	(h) $[FL/AT]$	(j) $[FL/A^2]$

2.26. $\text{lb}_m \cdot \text{ft/sec}^2$ (called a *poundal*)

2.27. $\text{kg}_f \cdot \text{s}^2/\text{m}$

2.28.
(a) $1 \text{ N} = 1 \text{ kg} \cdot \text{m/s}$ (force)		(h) $1 \, \Omega = 1 \text{ V/A}$ (electric resistance)
(b) $1 \text{ J} = 1 \text{ N} \cdot \text{m}$ (energy or work)		(i) $1 \text{ F} = 1 \text{ A} \cdot \text{s/V}$ (electric capacitance)
(c) $1 \text{ W} = 1 \text{ J/s}$ (power)		(j) $1 \text{ H} = 1 \text{ V} \cdot \text{s/A}$ (electric inductance)
(d) $1 \text{ Hz} = 1 \text{ s}^{-1}$ (frequency)		(k) $1 \text{ Wb} = 1 \text{ V} \cdot \text{s}$ (magnetic flux)
(e) $1 \text{ Pa} = 1 \text{ N/m}^2$ (pressure)		(l) $1 \text{ T} = 1 \text{ Wb/m}^2$ (magnetic flux density)
(f) $1 \text{ C} = 1 \text{ A} \cdot \text{s}$ (electric charge)		(m) $1 \text{ lm} = 1 \text{ cd} \cdot \text{sr}$ (luminous flux)
(g) $1 \text{ V} = 1 \text{ W/A}$ (electric potential)		(n) $1 \text{ lx} = 1 \text{ lm/m}^2$ (illumination)

2.29.
(a) 10^{-2}	(c) 10^3	(e) 10^{-6}	(g) 10^{-9}
(b) 10^{-1}	(d) 10^6	(f) 10^{-3}	(h) 10^{-12}

2.30.	0.054 N	**2.35.**	2778 W	**2.40.**	0.4 F
2.31.	4.5 m/s²	**2.36.**	3.41 °C	**2.41.**	$6 \times 10^7 \, \mu\text{s}$
2.32.	0.4079 kg	**2.37.**	25 V	**2.42.**	10^6 mg
2.33.	392.28 J	**2.38.**	1250 Ω	**2.43.**	10^9 ps
2.34.	0.0239 kcal/s	**2.39.**	1200 C	**2.44.**	10^6

2.45. *(a)* 5630.45 poundals   *(c)* 350 lb$_f$ = 11 260.9 poundals

　　　　(b) 175 lb$_m$ *(d)* 175 lb$_m$

2.46. 120.65 ft/sec^2 **2.48.** 417.083 lb$_m$ **2.50.** 82 500 ft · lb$_f$/sec   **2.52.** 74.1 °F

2.47. 298.38 ft · lb$_f$ **2.49.** 105 811 lb$_f$ **2.51.** 530.216 hp

2.53. *(a)* 344 lb$_m$　　　　*(h)* 2.36×10^{-4} ft　　*(o)* 52.4 kcal　　　　*(v)* 8.37 MW

　　　　(b) 21 400 g　　　　*(i)* 1.98×10^6 μs　　*(p)* 2.19×10^5 J　　*(w)* 4.49 gal

　　　　(c) 64.7 kg　　　　*(j)* 12×10^3 ps　　*(q)* 4.66×10^{-2} hp · hr　*(x)* 18.9 L

　　　　(d) 190 m　　　　*(k)* 1.61×10^{-5} h　　*(r)* 3.47×10^{-2} kW · hr　*(y)* 6167 m^3

　　　　(e) 15.4 in　　　　*(l)* 180 lb$_f$　　　　*(s)* 5811 kW · hr　　*(z)* 8.1×10^5 Pa

　　　　(f) 2.54×10^5 Å　　*(m)* 7.45×10^5 dynes　*(t)* 522 kW

　　　　(g) 322 km　　　　*(n)* 3.95 N　　　　*(u)* 0.570 hp

2.54. smaller

2.55. *(a)* 624.67 °R, 73.89 °C, 347.04 K　*(c)* 666.67 K, 740.33 °F, 393.52 °C

　　　　(b) 298.15 K, 77 °F, 536.67 °R　　*(d)* −53.15 °C, 396 °R, −63.67 °F

2.56. *(a)* 120 K, 216 °R　*(c)* 1800 °F, 1000 °C

　　　　(b) 25 °C, 25 K　　*(d)* 270 °F, 150 K

2.57. *(a)* 1.68 g/cm^3　*(d)* 7.7×10^6 J/min　*(g)* 5.651×10^{-5} kcal/s·cm^2　*(j)* 789 L/min

　　　　(b) 57.1 lb$_m$/ft^3　*(e)* 41 660 ft·lb$_f$/sec　*(h)* 53 N/cm　　　　*(k)* 85.17 W/m^2·K

　　　　(c) 31 kcal/s　*(f)* 1290 Btu/ft^2　*(i)* 0.332 kg/s·m^2　　*(l)* 1.72 stokeses

2.58. 25.6 N

2.59. 73.6 in^2

2.60. 0.125 ft/sec = 0.137 km/h

2.61. *(a)* consistent   *(b)* inconsistent   *(c)* consistent

2.62. c is dimensionless, provided $[F] = [ML/T^2]$ (as in SI)

2.63. *(a)* $[1/\theta]$ *(b)* K^{-1}

2.64. *(a)* $[ML^2/T^2\theta]$, $[L^3]$, $[ML^5/T^2]$. If force is considered a fundamental dimension, then the proper dimensions will be $[FL/\theta]$, $[L^3]$, $[FL^4]$.

　　　　(b) Pa · m^3/mol · K, m^3, Pa · m^6

　　　　(c) atm · ft^3/mol · °R, ft^3, atm · ft^6

2.65. *(a)* $[M/LT^2]$ or $[F/L^2]$ *(b)* N/m^2 or Pa   *(c)* in, lb$_f$/in^2

2.66. *(a)* $[M/T^3\theta]$ or $[F/LT\theta]$ *(b)* W/m^2 · K   *(c)* Btu/hr · ft^2 · °F or Btu/hr · ft^2 · °R

Chapter 3

3.10.   (*a*) 180 m   (*c*) $195 - 155 = 40$ m   (*e*) $6.4 - 5.5 = 0.9$ s   (*g*) 5.5 s
         (*b*) 78 m   (*d*) 4.5 s   (*f*) $5.2 - 3.5 = 1.7$ s

3.11.   (*a*) 1.73 mol/L
         (*b*) 2.05 mol/L, assuming the curve remains flat
         (*c*) 2.8 s
         (*d*) $5.8 - 2.6 = 3.2$ s
         (*e*) $1.24 - 1.20 = 0.04$ mol/L; $(0.04/1.24) \times 100 = 3.2$ percent, based upon the actual (measured) value

3.12.   (*a*) 100%, 65.5%, 7.5%   (*c*) 1268 K, 1305 K
         (*b*) 1248 K, 1288 K, 1326 K   (*d*) 1313 K, 62%

3.13.   (*a*) $(-2.6) - (-0.6) = -2.0$ N
         (*b*) $(-0.6) - (+0.6) = -1.2$ N
         (*c*) The minus sign indicates that the net change in force acts in the direction of decreasing displacement.

3.14.   (*a*) 360×10^7 N/m²   (*b*) 0.002 25   (*c*) $(450 \times 10^{-5})\left(\dfrac{0.001}{200 \times 10^7}\right) = 2.25 \times 10^{-15}$

3.15.   (*a*) 5.5%   (*b*) 89.9–90.1 mm

3.16.   (*a*) 7%   (*b*) 90.01 mm   (*c*) 89.71 mm

3.17.   (*a*) $y = 0.64$ (*c*) $x = 1.8$
         (*b*) $y = 0.1$ (*d*) $x = 6.45$ (*e*) $\dfrac{\log 0.1 - \log 10}{9.2 - 0} = -0.217$

3.18.   (*a*) 4.6×10^5 s⁻¹   (*c*) $306 - 324 = -18$ K
         (*b*) 336 K
                      (*d*) $\dfrac{\log (1 \times 10^5) - \log (1000 \times 10^5)}{0.003\,46 - 0.002\,90} = -5360$

3.19.   (*a*) 3   (*d*) 4.5
         (*b*) 1.05
         (*c*) 1.25   (*e*) $\dfrac{\log 10.44 - \log 0.2}{\log 9 - \log 1} = 1.8$

3.20.   (*a*) 114 °C, 86 °C, 114 °C, 86 °C   (*b*) 119 °C, 102 °C, 92 °C, 116 °C   (*c*) 105° and 255°

3.21.   (*a*) 55% Y, 45% X   (*c*) 75% Z, 25% Y   (*e*) 5% X, 85% Y, 10% Z
         (*b*) 90% X, 10% Z   (*d*) 15% X, 20% Y, 65% Z

3.22.   (*a*) 15% benzene, 65% acetone   (*c*) single-phase region
         (*b*) two-phase region   (*d*) 25% water, 12% benzene, 63% acetone

3.23.   (*a*) $71 - 48 = 23\%$ (*b*) $3.5 - 0.8 = 2.7\%$

3.24. $F = -30x$

3.25. $F = \begin{cases} -18x & x < 0 \\ -12x & x > 0 \end{cases}$

3.26. Relative maximum values are about 85 and 90 ppm, occurring at about 10:00 and 18:00 hours (10:00 A.M. and 6:00 P.M.), respectively.   The absolute minimum is 11.9 ppm at 4:00 (4:00 A.M.).   Other minima (relative minima) are 39.2 ppm and 28.7 ppm, occurring at 15:00 and 24:00 (3:00 P.M. and midnight).

3.28. (d) $v = 0.1(1 - e^{-0.2t})$

3.29. (c) $\mu = 1.1 \times 10^6\, T^{-1.5}$

3.30. $A = 6 \times 10^{12}\,\text{s}^{-1}$, $E = 80\,000$ J/mol

3.31. $c_1 = 0.1\ \text{h}^{-1}$, $c_2 = 0.015\ \text{h}^{-1}$

3.32. Plot $(T - 100)$ against t.

3.37. (c) The exam scores are not normally distributed, since the cumulative distribution cannot be represented by a straight line on probability paper.

3.38. (b) A semilog coordinate system, with the logarithmic scale along the abscissa.   Plot h versus $\log(1 + 8v)$.

3.40. $T = 50 + 30 \sin \theta$

3.41. (a) 1.52, 2.83, 0.23, 0.99 m   (b) 1.6, 6.9, 7.4, 5.5 ft

3.47. (a) 8.1   (b) 4.8   (c) 27.5   (d) 12.6

3.48. (a) −27   (b) 32   (c) 4.5   (d) −5.5

3.49. (a) 109   (b) 122   (c) 45   (d) 4   (e) 56

3.50. (a) 10.9   (b) 13.5   (c) 21   (d) 8.9

3.51. (a) 8.1 lb $SO_2/10^6$ Btu   (b) 1.4 % sulfur   (c) 8000 Btu/lb

Chapter 4

4.24. (a) ± 5 V   (b) 175 V   (c) 175 ± 5 V   (d) 340 ± 5 V   (e) 60 ± 5 V

4.25. (a) ± 2.5 m³/min
(b) 65 ± 2.5 m³/min.   (If the flow rate were recorded as 65.0 ± 2.5 m³/min, the last *two* digits would be uncertain.)
(c) 22.5 ± 2.5 m³/min or, rounding down, 22 ± 2.5 m³/min.
(d) 77.5 ± 2.5 m³/min or, rounding up, 78 ± 2.5 m³/min.

4.26. (*a*) Since the boiling point of pure water is 100 °C (at atmospheric pressure), it appears that the measurements are about 3 °C too high.   This is a consistent error.
(*b*) The scatter in the data indicates that random errors are present.
(*c*) Subtract 3 °C from each temperature, then plot the corrected data and pass a smooth curve through the data points.

4.27. (*a*) −3.01 °C
(*b*) −0.07, 0.11, 0.07, −0.04, 0.04, −0.13, −0.01 °C
(*c*) For this particular set of data the average will not be affected (the average of the five remaining points is still −3.01 °C).

4.28. (*a*) 11.9 cm
(*b*) 0.3, −0.2, 0.5, 0.1, −1.5, 0.0, 0.5 cm
(*c*) The fifth measurement will be rejected, resulting in a new average of 12.1 cm (determined from the remaining six measurements).
(*d*) 0.1, −0.4, 0.3, −0.1, −0.2, 0.3 cm

4.29. (*a*) 0.012 m/s^2 (*b*) 0.001 224   (*c*) 0.1224 %

4.30. (*a*) 29.13 mph   (*b*) 53.40 mph   (*c*) 77.67 mph

4.31. (*a*) −6.83 %   (*b*) $0.44

4.32. (*a*) 4.4 %   (*b*) 450–550 μF

4.33. (*a*) 14.98 m^3 (*b*) 1.976 %

4.34. (*a*) $\dfrac{2396 \text{ mi}}{103.9 \text{ gal}}$ = 23.06 mpg   (*b*) 21.96 mpg

4.35. 260.5 V

4.36. (*a*) 5.32 %   (*b*) 47.2 % high-sulfur coal, 52.8 % low-sulfur coal

4.37. 559.4 mmHg

4.38. 2.941

4.39. 3.352

4.40. 2.976

4.41. (*a*) $\bar{x} = 8.0$, $\tilde{x} = 8.0$, mode = 8.2, $v = 0.125$, $s = 0.354$
(*b*) $\bar{x} = 139.25$, $\tilde{x} = 139$, $v = 172.94$, $s = 13.15$
(*c*) $\bar{x} = 65.74$; $\tilde{x} = 65.4$; modes = 65.4, 67.2; $v = 1.262$; $s = 1.124$
(*d*) $\bar{x} = 0.53$, $\tilde{x} = 0.8$, $v = 12.1$, $s = 3.5$

4.42. $\bar{x}_A = 3669.2$ N, $s_A = 181.6$ N; $\bar{x}_B = 3595.7$ N, $s_B = 252.2$ N.   The group A data appears more accurate, because its standard deviation is smaller.

4.43. $24 018

4.44.   10

4.45.   See Table A-1, columns 1–3.

Table A-1

Interval	Number of Data Points	Relative Frequency,%	Cumulative Relative Frequency,%
0.01–5	0	0	0
5.01–10	1	2	2
10.01–15	1	2	4
15.01–20	1	2	6
20.01–25	1	2	8
25.01–30	1	2	10
30.01–35	0	0	10
35.01–40	1	2	12
40.01–45	4	8	20
45.01–50	5	10	30
50.01–55	2	4	34
55.01–60	3	6	40
60.01–65	4	8	48
65.01–70	6	12	60
70.01–75	1	2	62
75.01–80	7	14	76
80.01–85	3	6	82
85.01–90	3	6	88
90.01–95	2	4	92
95.01–100	4	8	100
TOTALS	50	100%	

4.46.   (a) $\bar{x} = 63.54$, $s = 22.67$ (b) $\bar{x} = 63.30$, $s = 22.61$

4.47.   See Table A-1, column 4.

4.48. $\tilde{x} = 65.84$

4.49. $\bar{x} = 4.98$, $\tilde{x} = 4.50$, $s = 1.94$

4.50. $\bar{x} = 90.00$ mm, $\tilde{x} = 90.00$ mm, $s = 0.2545$ mm

4.51.   (a) 12 students   (b) 24 students   (c) $62 - 40 = 22$ students

4.52.   (a) 16%   (c) 90%   (e) $90 - 16 = 74\%$
            (b) $100 - 16 = 84\%$ (d) 10%   (f) $66 - 46 = 20\%$

4.53.   (a) 6.2 ppm   (b) 1 ppm or less

4.54.   (a)   14.5%   (b)   13.5 + 4.5   =   18.0%   (c)   9.0%   (d)   100.0 − 9.0   =   91.0%
(e)   1.5%

4.55.   (a) $\bar{x} = 10.1$, $s = 3.4$ (b) $\bar{x} = 4.45$, $s = 1.90$

4.56.   The bell-shaped normal curve better conveys the character of the normal distribution, but the straight-line plot on probability paper is easier to construct and to read.

4.57. $\mu = 62$, $\sigma = 25$

4.58.   The median and the mode will be the same as the mean, whose value is approximately 62.

4.59.   (a)   The curve would be shifted to the right (for an increase in μ) or the left (for a decrease).
(b)   The curve would become shorter and broader (for an increase in σ) or taller and narrower (for a decrease).

4.60. $F = -1.667 + 0.667x$

4.61. $S = -6.063\,38 + 5.264\,41\rho$

4.62. $y = 0.80 - 187.5x$

4.63   (a) $v = 1.498\,53\,r^{0.801\,85}$ (b) $\mu = 0.947\,919 \times 10^6 T^{-1.474\,82}$

4.64.   (a) $k = 3.637\,27 \times 10^{23} e^{-12\,364/T}$ (b) $d = 0.100\,17\,e^{2.999\,23\,t}$ (c) $k = 6.612\,87 \times 10^{12} e^{-9658/T}$

4.65. $R = 1000 - 999.82\,e^{-0.079\,98\,x}$

4.66.   (a) $E = 2.104\,36 \times 10^{-3} T^{2.325\,46}$
(b) $E = 0.261\,197 - 0.146\,182\,T + 0.016\,563\,5\,T^2$
(c)   The quadratic function provides the best fit.

4.68.   (a) $r = 0.992\,738$.   The data can be represented accurately by a straight line having a positive slope.
(b) $r = -0.310\,370$.   The data cannot be accurately represented by a straight line.
(c) $r = -0.997\,776$.   The data can be represented accurately by a straight line having a negative slope.

Chapter 5

5.16.   (a) $x = 1.36$ (c) $x = 2.00$ (e) $x = 1.16$
(b) $x = 1.57$ (d) $x = 1.25$ (f) $x = 2.41$

5.17. $x = -2.2$, $x = 1.4$, $x = 3.6$

5.19.   (a) $x = 1.351\,96$ (e) $x = 1.165\,56$ (h) $x = -1.718\,960$
(b) $x = 1.568\,95$ (f) $x = 2.412\,01$ $x = 1.951\,343$
(c) $x = 2.000\,00$ (g) $x = -0.604\,00$ $x = 14.434\,284$
(d) $x = 1.250\,00$

5.24. (a) 13 iterations   (b) 9.2×10^{-5}, 4 decimals

5.25. (a) $t = = 0.534$ ms
       (b) $v_{max} = 1.490$ mV, $t = 2.747$ ms
       (c) $v_{min} = -0.795$ mV, $t = 9.03$ ms

5.26. $T = 4\,°C$, $V_{min} = 0.9998 \times 10^{-3}$ m³/kg

5.27. $R = 0.1413$ m

5.28. (a) $x = 2.218$ (b) $R = 11.09$ mm

5.29. (a) 0.007 083 33 (0.71 % per month)   (b) 0.085 (8.5 % per year)   (c) \$110 725

5.30. (a) $x_1 = 3$, $x_2 = -1$, $x_3 = 4$
       (b) $x_1 = -2$, $x_2 = 2$, $x_3 = 1$
       (c) $x_1 = 0.5$, $x_2 = -1.5$, $x_3 = 2.5$, $x_4 = -2$
       (d) $x_1 = 2.9792$, $x_2 = 2.2156$, $x_3 = 0.2113$, $x_4 = 0.1523$, $x_5 = 5.7150$

5.35. 28.788 tons Al/day, 6.291 tons A2/day, 25.349 tons A3/day, 44.686 tons A4/day

5.36. (a) One set is:

$$I_1 + I_2 + I_3 = 2 \qquad\qquad 10I_1 - 40I_2 + 10I_4 = 0$$
$$I_1 - I_4 + I_6 = 0 \qquad\qquad 10I_4 + 2I_6 - 50I_7 = 0$$
$$I_6 + I_7 + I_{10} = 3 \qquad\qquad 50I_7 - 10I_8 - 5I_{10} = 0$$
$$I_8 - I_9 - I_{10} = 0 \qquad\qquad 20I_5 - 10I_8 - 10I_9 = 0$$
$$I_3 - I_5 - I_9 = 0 \qquad\qquad 40I_2 - 5I_3 - 20I_5 = 0$$

       (b) $I_1 = 0.406$ A $I_3 = 1.048$ A $I_5 = 0.829$ A $I_7 = 0.410$ A $I_9 = 0.219$ A
           $I_2 = 0.546$ A $I_4 = 1.776$ A $I_6 = 1.369$ A $I_8 = 1.439$ A $I_{10} = 1.221$ A

5.37. $T_0 = 197.0\,°C$, $T_1 = 49.3\,°C$, $T_2 = 43.4\,°C$, $T_3 = 23.7\,°C$

5.39. (a) 131.3 km

5.43. (a) 0.693   (b) 3.14   (c) 0.479   (d) −0.747

5.46. (a) 1.250   (b) 1.330

5.47. 1.333 333

5.48. (a) 1.333 333   (b) 1.333 333

5.49. (a) 1   (b) 0.159 777   (c) −0.186 633   (d) 2/3

5.50. (a) 37 860 bbl   (b) 105.2 bbl/day

5.51. (a) 40.4 ppm   (b) 50.9 ppm

5.52.   (*a*) 95.7 s   (*b*) 98.7 s   (*c*) 101.7 s   (*d*) $\dfrac{30}{101.7} = 0.295$ km/s   (*e*) 2.58 km/s
(Answers obtained using 2 km intervals.)

5.53.   653.3 °C

5.54.   55.04 mV

5.55. 0.4363×10^{-3} m³

Chapter 6

6.10.   (*a*) 2.2500   (*c*) 2.0125   (*e*) 3.5000
(*b*) 1.9000   (*d*) 1.6600   (*f*) 1.5625

6.11.   (*a*) 3.3864   (*c*) 2.6639   (*e*) 10.5451
(*b*) 2.3818   (*d*) 1.8704   (*f*) 1.7285

6.12.   (*a*) $3500   (*b*) approximately 7% per year

6.13.   Simple interest increases linearly, compound interest increases exponentially.

6.14.   (*a*) $13 266.49   (*b*) $13 507.42   (*c*) $16 035.68   (*d*) $16 453.31

6.15.   (*a*) $4006.41   (*b*) $4006.95

6.16. $4045.80 (an increase of almost $40)

6.18.   8% per year

6.19.   (*a*) $21 459.35   (*b*) $22 160.23   (*c*) $22 324.85   (*d*) $22 408.44
Continuous compounding yields almost $1000 more than annual compounding.   However, the difference between continuous compounding and monthly compounding is small.

6.20.   11 years

6.21.   (*a*) 47.7271   (*c*) 24.6504   (*e*) 76.3608
(*b*) 18.4237   (*d*) 7.9129   (*f*) 12.9510

6.22.   (*a*) $79 058.19   (*b*) 33 years

6.23. $1264.89

6.24. $\dfrac{\$1264.89}{1.06} = \$1193.29$

6.25.   62.4 years

6.26.   (*a*)   0.0710   (*c*)   0.1081   (*e*)   0.1381
         (*b*)   0.1293   (*d*)   0.2364   (*f*)   0.1335

6.27.   (*a*) $15 260
         (*b*)   12.8 years
         (*c*)   The account will never be depleted, because only the annual interest is withdrawn each year (the principal remains intact).

6.28.   10.6% per year

6.29.   (*a*) $626.25   (*b*) $631.25

6.30.   (*a*)   The lump-sum payment, deposited in the bank, will produce equal annual payments of $54 391.56 for 20 years.   Therefore it would be wise to accept the lump-sum payment.
         (*b*)   No, since taxes are likely to reduce significantly the amount of money that can be deposited in the bank.

6.31. $A = (1 + ni)P/12n$

6.32.   The "add-on" loan would cost $160.00 per month, and the conventional loan would cost $158.00 per month.   Thus the conventional loan will be somewhat less expensive, despite the higher annual interest rate.

6.33.   (*a*) $523.40
         (*b*) $157 020, including $92 020 in interest (note that the interest is substantially more than the original loan).

6.34.   (*a*) $164 324   (*b*) $350 271   (*c*) $312 667

6.35.   (*a*) $300 000   (*b*) $109 637   (*c*)   −$8 613   (*d*)   −$110 304

6.36.   The present worth will have a value of zero at an annual interest rate of approximately 6.8%.   At this interest rate, the expected future revenues will be just offset by the initial expenditure.

6.37.   (*a*) $300 000   (*b*) $138 790   (*c*)   −$13 027   (*d*)   −$195 325

6.38.   6.8% (as in Problem 6.36)

6.39. $P = $287 235 $F = $634 820

6.40.   (*a*) $P_A = $250 000   (*b*) $P_A = $149 351   (*c*) $P_A = $68 613
               $P_B = $320 000 $P_B = $175 444 $P_B = $63 410
             Choose investment B   Choose investment B   Choose investment A

6.41. $P_A = $23 558$, $P_B = 5042;   hence investment A is more desirable.   If the time value of money is not considered, then $P_A = P_B = $350 000$.   Thus the investments would be equally desirable.

6.42 $P_1 = $9 050 000$ for the midwestern plant, $P_2 = $11 624 000$ for the west coast plant; build the west coast plant.

6.43. $P_1 = \$9\,050\,000$ for the midwestern plant, $P_2 = \$7\,118\,000$ for the west coast plant; expand the midwestern plant.

6.44. $P_1 = \$5\,000\,000$, $P_2 = \$5\,036\,000$;   first proposal is better.

6.45. $P_1 = \$5\,000\,000$, $P_2 = \$4\,851\,537$;   second proposal is better.

Chapter 7

7.8.   See Fig. A-6.

7.9.   See Fig. A-7.

7.10.   (*b*) See Fig. A-8.

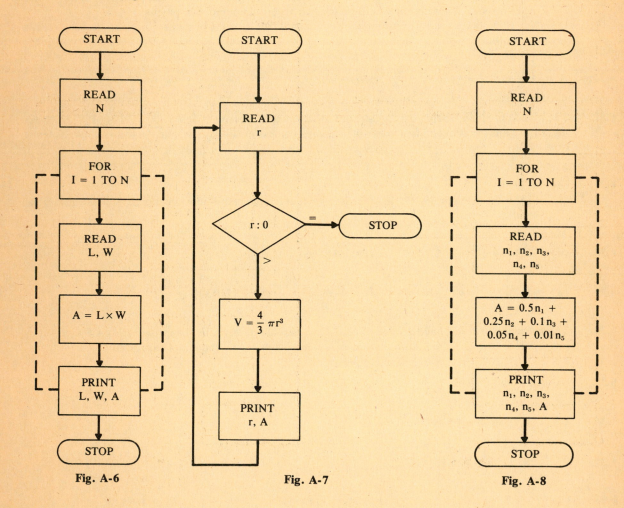

Fig. A-6   **Fig. A-7**   **Fig. A-8**

7.11   (*c*) See Fig. A-9.

7.12.   See Fig. A-10.

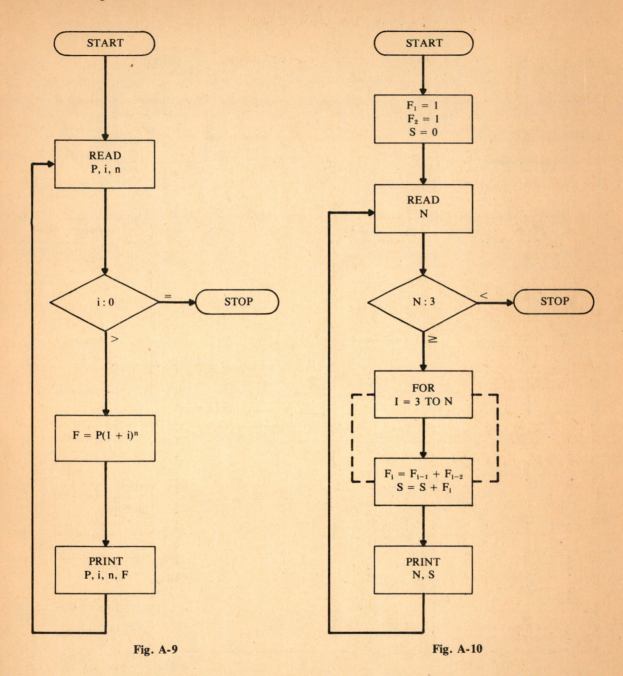

Fig. A-9   Fig. A-10

7.15. See Fig. A-11.

7.17. See Fig. A-12.

7.18. See Fig. A-13.

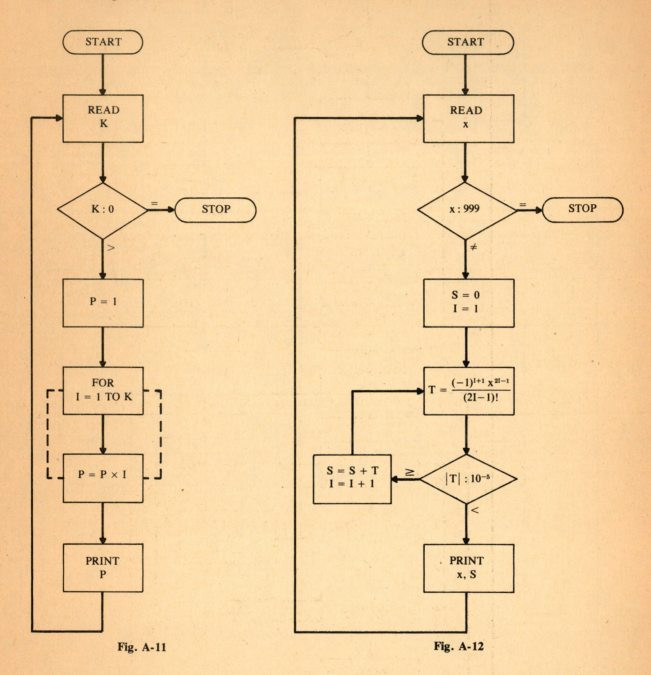

Fig. A-11　　　　　　　　　　　　　Fig. A-12

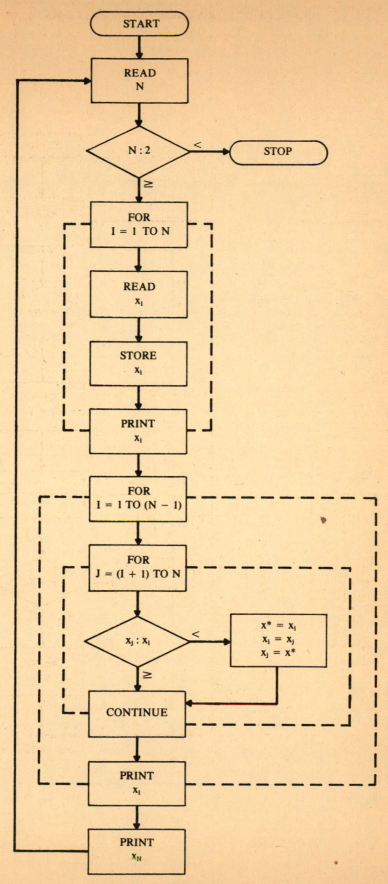

Fig. A-13

7.19.   See Fig. A-14.

7.21.   See Fig. A-15.

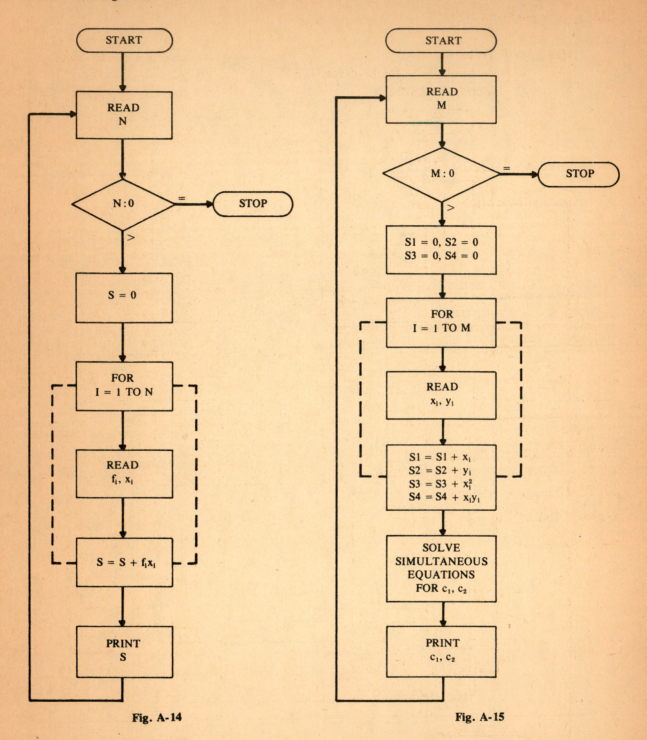

Fig. A-14   Fig. A-15

7.24.   See Fig. A-16.

7.25.   (*a*) See Fig. A-17.

7.26.   (*a*) See Fig. A-18.

Let N = number of courses per term
       C_i = number of credits for the *i*th course
       G_i = grade for the *i*th course

Let N = maximum number of iterations
       x_0 = initial guess
       ϵ = convergence criterion

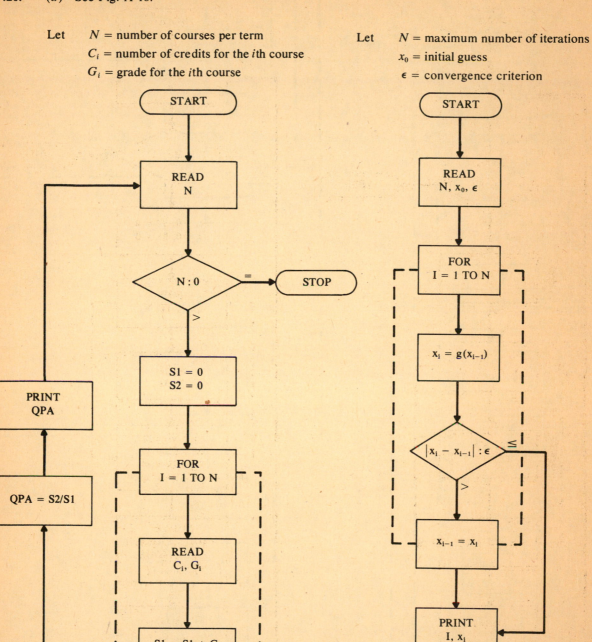

Fig. A-16   Fig. A-17

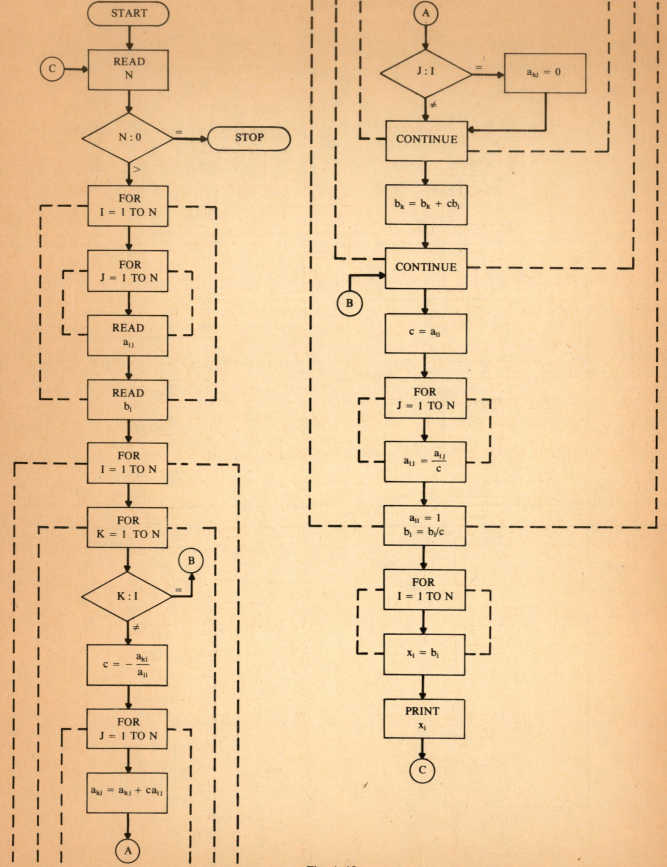

Fig. A-18

7.27. See Fig. A-19.

7.28. (*b*) See Fig. A-20.

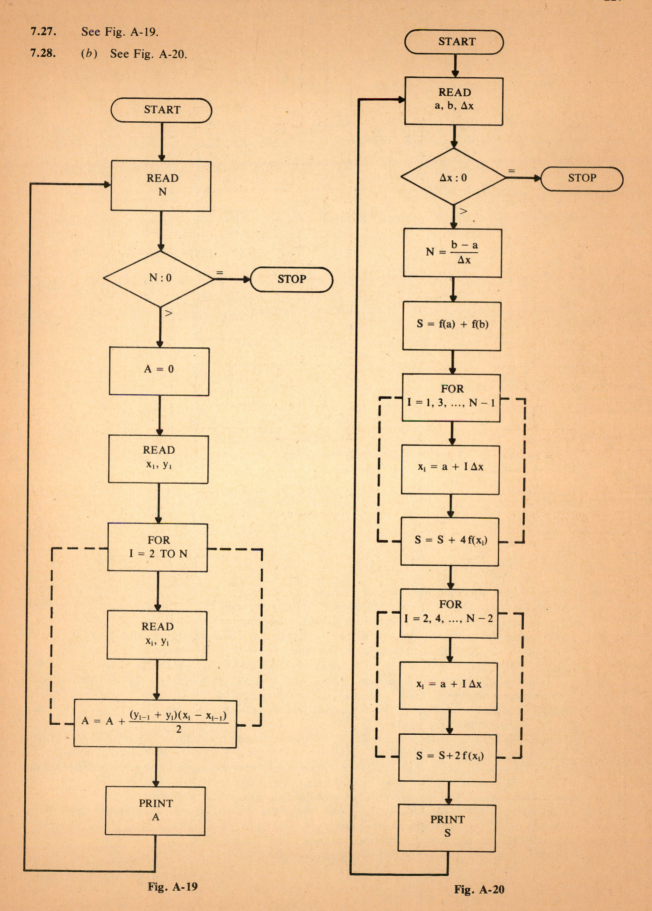

Fig. A-19

Fig. A-20

7.29.   See Fig. A-21.

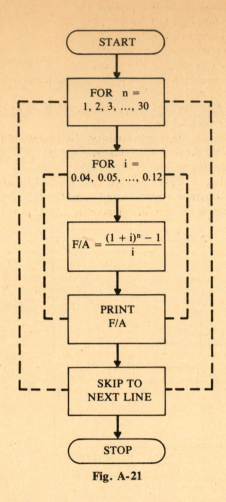

Fig. A-21

7.30.   First two lines are comments (program heading).
Next 8 lines (100 FORMAT ... 700 FORMAT): statements which specify the appearance of the input and output data, and provide English labels for the output.
Next 4 lines (READ, DO 5, READ, WRITE): read input data and then print values for a, b, c.
Next line (IF ... GO TO 4): skips to statement number 4 (X= $-$C/B) if $a = 0$.
Next 2 lines (DISC =, IF): skip to statement number 3 (XREAL = ...) if the *discriminant* ($b^2 - 4ac$) is negative, to statement number 2 (X = $-$ B/(2.*A)) if the discriminant equals zero, and to statement number 1 (X1 = ...) if the discriminant is positive.
Next 7 lines (comments through GO TO 5): calculate the real roots (X1 and X2), print the results, and skip to the end of the program.
Next 6 lines (comments through GO TO 5): calculate a single repeated root (X), print the result, and skip to the end of the program.
Next 7 lines (comments through GO TO 5): calculate the real and imaginary components (XREAL and XIMAG) of the complex roots, print the results, and skip to the end of the program.
Next 4 lines (comments through WRITE): calculate a single root (X) and print the result.
Next line (5 CONTINUE): this is a dummy statement that provides an entry point for the above skips.
Next line (STOP): terminates program execution.
Last line (END): identifies the physical end (last line) of the program.
(Note that this program is oriented toward batch processing.)

7.31. Lines 10–20 are remarks (program heading).

Line 30: prints a message requesting input data.

Line 40: enters input data into computer.

Line 50: tests for stopping condition.

Line 60: skips to line number 430 if $a = 0$.

Line 70: evaluates the *discriminant* ($b^2 - 4ac$).

Lines 80–90: skip to line number 330 if the discriminant is negative, and to line number 230 if the discriminant equals zero.

Lines 100–160: calculate the real roots (X1 and X2), print the results, and return to line 30 (thus beginning another calculation).

Lines 200–250: calculate a single repeated root (X), print the result, and return to line 30.

Lines 300–360: calculate the real and imaginary components (R and I) of the complex roots, print the results, and return to line 30.

Lines 400–450: calculate a single root (X), print the result, and return to line 30.

Line 500: terminates program execution and identifies the end (last line) of the program.

(Note that this program is written for timesharing applications.)

Index

Catalog

If you are interested in a list of SCHAUM'S
OUTLINE SERIES send your name
and address, requesting your free catalog, to:

SCHAUM'S OUTLINE SERIES, Dept. C
McGRAW-HILL BOOK COMPANY
1221 Avenue of Americas
New York, N.Y. 10020